Impacts on biodiversity of exploitation of renewable energy sources: the example of birds and bats - facts, gaps in knowledge, demands for further research, and ornithological guidelines for the development of renewable energy exploitation

Hermann Hötker, Kai-Michael Thomsen, Heike Jeromin

Translated by: Solveigh Lass-Evans

Michael-Otto-Institut

Supported by:

Imprint

Unchanged traslation of: Hötker, H.; Thomsen, K.-M. & H. Köster (2005): Auswirkungen regenerativer Energiegewinnung auf die biologische Vielfalt am Beispiel der Vögel und der Fledermäuse - Fakten, Wissenslücken, Anforderungen an die Forschung, ornitologische Kriterion zum Ausbau von regnerativen Energiegewinnungsformen. BFN-Skripten 142, Bonn.

Recommended citation:
Hötker, H., Thomsen, K.-M. & H. Jeromin (2006): Impacts on biodiversity of exploitation of renewable energy sources: the example of birds and bats - facts, gaps in knowledge, demands for further research, and ornithological guidelines for the development of renewable energy exploitation. Michael-Otto-Institut im NABU, Bergenhusen.

Translation:
Solveigh Lass-Evans

Layout:
Kai-Michael Thomsen

Photographs:
Dr. Hermann Hötker, Kai-Michael Thomsen

Supported by:
Bundesamt für Naturschutz (Support No. Z1.3-684 11-5/03)
and Bundesministerium für Umwelt, Naturschutz und Reaktorsicherheit (Translation)

Editor:
Michael-Otto-Institut im NABU, Goosstroot 1, D-24861 Bergenhusen

Published by:
Books on Demand GmbH, Gutenbergring 53, D-22848 Norderstedt

ISBN 3-8334-5257-9

Table

Preface 6

Summary 7

1 Introduction 9

2 Material and Methods 11

3 Impacts of wind farms on vertebrates 15
3.1 Non-lethal impacts (disturbance, displacement, habitat loss) on birds..... 15
3.1.1 Change in distributions due to wind farms 15
3.1.2 Minimum avoidance distance of birds to windfarms 15
3.1.3 Barrier effects of wind farms on birds 25
3.3 Collision of birds and bats with wind farms 28
3.3.1 Collisions of birds with wind farms 28
3.3.2 Collisions of bats with wind farms 34

4 Collison effects on population dynamics 37
4.1 Application of the population simulation model 37
4.2 Results of the population simulations 40
4.3 Discusion of the simulations calculations 40

5 Measures to reduce the impacts of wind farms 43
5.1 Coice of site 43
5.2 Design of the environment around wind farms 43
5.3 Configuration of wind turbines within a wind farm 44
5.4 The operation of wind farms 44
5.5 Design of individual wind farms 44
5.6 Measures transferred on the conditions/circumstances in Germany 45

6 Estimation of the impacts of 'repowering' 46
6.1 Repowering and disturbance of birds 46
6.2 Repowering and collisions of birds and bats 48
6.3 Summary of the assessment of repowering 48

7 Impacts of other types of renewable energy 50

8 Research requirements 52

9 Acknoledgements 55

10 Literature 56

Preface

The Federal Government of Germany aims to increase the proportion of energy generated by renewable sources to at least 12.5% by 2010. The regulations of the „renewable-energy-bill" (EEG) offer incentives to do this. For the first time, in the first half of 2004, 10% of the power supply in Germany was generated from renewable sources and a substantial proportion (5.8% of total) is provided by wind energy. At the same time, the associated impacts on ecological balance and landscape caused by wind farms construction is a controversial topic.

The aim of this study „Impacts on biodiversity of exploitation of renewable energy sources: the example of birds and bats - facts, gaps in knowledge, demands for further research, and ornithological guidelines for the development of renewable energy exploitation" compiled by the Michael-Otto-Institute in NABU, is to analyse national and international studies of the impacts of wind farms on birds and bats and to access their statistical significance. The results give an overview of the available data, as well as gaps in knowledge and future research needs.

The publications used in this study allow conclusions to be drawn on the impacts wind farms have on breeding, roosting, foraging and migrating birds and bats. Furthermore, some conclusions can be reached about the impacts of wind farms on these taxa throughout the year. Simulation models assess the additional mortality rate caused by wind farms and the effects of repowering on the bird and bat populations.

However, note that the data are limited and open to a wide range of interpretation and therefore only tentative conclusions can be reached on the impacts of wind farms on birds and bats.

The results presented here should objectivise the nation-wide discussions about the extent to which nature conservation is compatible with renewable energy development (in particular with wind farms). While the report was in progress, the results have been already discussed with administrative bodies, with representatives of nature conservation associations as well as the operating companies.

Prof. Dr. Hartmut Vogtmann
President of the Federal Nature Conservation Office

Summary

Impacts on biodiversity of exploitation of renewable energy sources: the example of birds and bats – facts, gaps in knowledge, demands for further research, and ornithological guidelines for the development of renewable energy exploitation

The purpose of this report is to compile and to evaluate the available information on the impacts of exploitation of renewable energy sources on birds and bats. The focus is on wind energy as there is only little information on the impact on birds and bats of other sources of renewable energy. The report aims at better understanding the size of the impact, the potential effects of re-powering (exchanging small old wind turbines by new big turbines), and possible measures to reduce the negative impact on birds by wind turbines. In addition the need for further research is highlighted.

The evaluation is based on 127 separate studies (wind farms) in ten countries, most of them in Germany. Most studies were brief (not more than two years) and did not include the pre-construction period. Before-After Control Impact studies that combine data collection before and after, in this case construction of a wind farm, on both the proposed development site and at least one reference site were rare. In only a few cases, would the design of the study and the length of the study period theoretically allow statistically significant effects of wind farms on birds and bats to be found at all. Assessments of impacts, therefore, are usually based on few studies only. This report includes all studies readily available to the authors, irrespective of the length of the study period and the quality of the study design. In order to base the assessments on as many independent samples as possible even rather unsystematic observations were included. The information of the data was reduced to a level that justified the application of sign tests. The compilation of many different individual studies gave the following results:

The main potential hazards to birds and bats from wind farms are disturbance leading to displacement or exclusion and collision mortality. Although there is a high degree of agreement among experts that wind farms may have negative impacts on bird populations no statistically significant evidence of negative impacts on populations of breeding birds could be found. There was a tendency waders nesting on open grounds to be displaced by wind farms. Some passerines obviously profited from wind farms. They were probably affected by secondary impacts, e.g. changes in land management or abandonment from agricultural use next to the wind plants.

The impact of wind farms on non-breeding birds was stronger. Wind farms had significantly negative effects on local populations of geese, Wigeons, Golden Plovers and Lapwings.

With the exceptions of Lapwings, Black-tailed Godwits and Redshanks most bird species used the space close to wind turbines during the breeding season. The minimal distances observed between birds and pylons rarely exceeded 100 m during the breeding season. Some passerines showed a tendency to settle closer to bigger than to smaller wind turbines.

During the non-breeding season many bird species of open landscapes avoided approaching wind parks closer than a few hundred metres. This particularly held true for geese and waders. In accordance with published information disturbance of geese may occur at least up to 500 m from wind turbines. For most species during the non-breeding season, the distances at which disturbance could be noted increased with the size of the wind turbines. For Lapwings this relationship was statistically significant.

There was no evidence that birds generally became „habituated" to wind farms in the years after their construction. The results of the few studies lasting longer than one season revealed about as many cases of birds occurring closer to wind farms (indications for the existence of habituation) over the years as those of birds occurring further away from wind farms (indications for the lack of habituation).

The question whether wind farms act as barriers to movement of birds has so far received relatively little systematic scientific attention. Wind farms are thought to be barriers if birds approaching them change their flight direction, both on migration or during other regular flights.

There is evidence for the occurrence of a barrier effect in 81 bird species. Geese, Common Cranes, waders and small passerines were affected in particular. However, the extent to which the disturbances due to wind farms of migrating or flying birds influences energy budgets or the timing of migration of birds remains unknown.

Collision rates (annual number of killed individuals per turbine) have only rarely been studied with appropriate methods (e. g. with controls of scavenger activities). In particular, such studies are missing for Germany. Collision rates varied between 0 and more than 50 collisions per turbine per year for both birds and bats. Obviously the habitat influenced the number of collisions. Birds were at high risks at wind farms close to wetlands where gulls were the most common victims and at wind farms on mountain ridges (USA, Spain), where many raptors were killed. Wind farms in or close to forests posed high collision risks for bats. For both birds and bats, the collision risk increased with increasing size of the wind turbine. The relationship, however, was not statistically significant.

Gulls and raptors accounted for most of the victims. In Germany the relatively high numbers of White-tailed Eagles (13) and Red Kites (41) killed give cause for concern. Germany hosts about half of the world population of breeding Red Kites and has a particular responsibility for this species. Bird species that were easily disturbed by wind farms (geese, waders) were only rarely found among the victims. Bats were struck by wind turbines mostly in late summer or autumn during the period of migration and dispersal.

Population models created by the software VORTEX revealed that significant decreases in size of bird and bat populations may be caused by relatively small (0,1 %) additive increases in annual mortality rates, provided they are not counter acted by density dependent increases in reproduction rates. Short-lived species with high reproductive rates are more affected than long-lived species with low reproductive rates. The latter, however, are less able to substitute additional mortality by increasing reproductive rates.

The effects of „repowering" (substitution of old, small turbines by new turbines with higher capacity) on birds and bats is assessed by the available data and by simple models. There is no information, however, on the effects of the newest generation of very large wind turbines. According to current knowledge, repowering will reduce negative impacts on birds and bats (disturbance and mortality) if the total capacity of a wind farm is not changed (many small turbines are replaced by few big turbines). In a scenario in which the capacity of a wind farm is increased 1.5 fold, negative impacts start to dominate. In case of a doubling of wind farm capacity, repowering increases the negative impacts of the wind farm. Repowering offers the chance to remove wind farms from sites that are associated with high impacts or risks for birds and bats. New turbines could be constructed on sites that are likely to be less problematic with respect to birds and bats.

Effective methods of reducing the negative impacts of wind energy use on birds and bats include:

- choice of the right site for wind farms (avoidance of wetands, woodlands, important sites for sensitive non-breeding birds and mountain ridges with high numbers of raptors and vultures),
- measures to reduce the attractivness of wind farm sites for potential collision victims,
- configuration of turbines within wind farms (placement of turbines parallel to and not accross the main migration or flight directions of birds),
- construction of wind turbines: replacement of lattice towers, wire-cables and overhead power lines.

Measures to increase the visibility of wind turbines and to reduce the effects of illumination remain to be studied.

In spite of many publications on windfarms and birds there still is a great demand for further research. First of all there is an urgent need for reliable data on collision rates at wind turbines of birds and bats in Germany. This holds true particularly for the new and big turbines which will replace the present generation of wind turbines.

It is still unclear whether these big and necessarily illuminated turbines pose a high collision risk to nocturnal migrants which have not yet been greatly affected by smaller turbines. The high collision rates of Red Kites in Germany also merit urgently study. The aim of the research has to be a quick reduction of the collision rates. The sensitivity to wind farms of many other bird species that are of particular nature conservation interest (storks, raptors, Cranes) has not yet been sufficiently studied.

There is hardly any information on the impacts of arrays of solar panels on birds and bats. Studies should be initiated as soon as possible.

1 Introduction

The Federal Government of Germany intends to set targets to develop further renewable energy generation (BWU, 2004) in order to minimise emissions of gases with damaging effects on climate. Germany generates more electricity from wind than any country, followed by the USA (BIOCENOSE & LPO Aveyron- Grands Causses, 2002; BMU, 2004). Even during the early phase of wind energy development, worries were expressed mainly in the USA, but also in Europe, that wind farms could have a harmful impact on the animal world, particularly on birds. In the USA, these apprehensions were confirmed by experiences from the first large-scale wind farm, where at the Altamont Pass in California, roughly 5000 wind turbines have been responsible for the death of hundreds of raptors per year since their construction. Protected species such as golden eagles are among the species affected (Orloff & Flannery, 1992). Due to high mortality of raptors, there has been no further extension of this wind farm and overall the development of wind energy in the USA has slowed down (Hunt, 2000). In contrast to the USA, discussions in Europe focused on more indirect impacts, such as disturbance, habitat loss during the breeding season and on migration, as well as barrier effects for migrating birds (AG Eingriffsregelung, 1996; Crockford, 1992; Langston & Pullan, 2003; Percival, 2000; Reichenbach, 2003; Schreiber, 2000; Winkelman, 1992b). Collisions of birds with wind turbines were also significant in Spain, where griffon vultures were particularly affected (Acha, 1998; Lekuona, 2001; SEO, 1995). Meanwhile, losses of raptors were also recorded at wind farms in Germany (Dürr, 2003a; Dürr, 2004). New studies in America and Europe show that bats as well as birds are be killed by wind farms (Ahlén, 2002; Bach, 2001; Dürr & Bach, 2004; Johnson et al., 2000; Keeley, Ugoretz & Strickland, 2001).

Despite numerous studies the extent of ecological impacts of wind farms is controversial. One possible reason might be the heterogeneity of the available studies. They differ in recorded parameters (number of killed and injured birds, disturbance), in their analytical design (collection of data before or after construction, use of control areas), in their extent, duration and in their applied analytical methods. Many studies are either published in barely assessable places, or not published at all („grey" literature). In only a few cases have studies been validated by independent experts (as is common in peer-reviewed journals). Therefore, in these circumstances, many specific questions could not be answered concretely or scientifically. It has not been possible to forecast what sort of impacts a wind farm in specific location might have on birds or bats. The same applies to other forms of renewable energy sources, particularly as considerably fewer research results exist.

Popular discussions in wind farm energy are not based on objective facts but rather on assumptions. However, informed debate is required for a popular consensus about the usage of wind energy. Along with the database developed at the same time, this study should help to objectivise these discussions.

This report attempts to summarise clearly the contemporaneous datasets that might be classified as „effects of renewable energy development on birds and bats". The main emphasis was on wind energy, as other forms of renewable energy (with the exception of hydroelectric) are not so well established and therefore relatively few observations have been made (see chapter 7).

The statements made in this report are based on bibliographic references and interviews with experts. The documented results of many studies were not only summarised, but also analysed. In this way, a wide range of observations and data on this topic were available to answer questions, which individual analyses would not be capable of solving. The numerous studies on the impact of wind energy enabled a search for connections, which identify, explain and maybe even help to diminish the negative impacts. The analysis of the data presented here from reports and publications about wind power should contribute

- to assess better the extent of impacts (if possible at the level of populations of species)
- to estimate the potential impact of repowering (replacement of the old, small wind turbines by bigger, more powerful (but possibly fewer) turbines)
- to identify possible measures to reduce negative impacts of wind farms on birds and bats.

The following questions have been considered in detail:

- do wind farms cause changes in the distribution of animal populations?
- how big is the distance animals keep to wind farms?
- do wind farms have a barrier effect on migrating birds?
- to what extent are wind farms responsible for mortality and what are the implications of the losses for population dynamics?
- how can the negative impacts of wind farms be mitigated?

The questions have been as far as possible treated for both birds and bats. In order to consider the consequences of mortality on wind farms, populations have been modelled with help of a computer simulation program. Because extensive studies about possible impacts of offshore-wind energy usage on the marine environment are at present still in progress (Kutscher, 2002), this report is restricted mainly to land and coastal areas The literature study was focused on central Europe, while from other areas, only comprehensive published studies were included.

2 Materials and Methods

The database and the presented results are both based exclusively either on published or „grey literature" reports. No new data were collected for this project. Within the project more than 450 literature references have been analysed, i.e. checked and if necessary added to the tables. Cross-checking showed that these references are derived from 127 different studies. Data from each wind farm was treated as a single study, even if the data was gathered in different years and from different observers. This was done in order to ensure the independence of the data and to avoid using the same study more than once. The following sources were used to derive the baseline data for this study:

Ahlén, 2002; Albouy et al., 1997; Albouy, Dubois & Picq, 2001; Anderson et al., 2000b; Bach, in Druck ; Bach, 2001; Bach, 2002; Bach, Handke & Sinning, 1999; Barrios & Rodriguez, 2004; Bergen, 2001a; Bergen, 2001b; Bergen, 2002a; Bergen, 2002b; Bergh, Spaans & Swelm, 2002; Boone, 2003; Böttger et al., 1990; Brauneis, 1999; Brauneis, 2000; Clemens & Lammen, 1995; De Lucas, Janss & Ferrer, 2004; Dulas Engineering Ltd, 1995; EAS, 1997; Erickson, Kronner & Gritski, 2003; Everaert, 2003; Everaert, Devos & Kuijken, 2002; Förster, 2003; Gerjets, 1999; Gharadjedaghi & Ehrlinger, 2001; Guillemette & Larsen, 2002; Guillemette, Larsen & Clausanger, 1999; Hall & Richards, 1962; Hormann, 2000; Hydro Tasmania, ; Isselbächer & Isselbächer, 2001; Janss, 2000; Johnson, 2002; Johnson et al., 2003; Johnson et al., 2000; Kaatz, 2000; Kaatz, 2002; Kerlinger, 2000; Ketzenberg et al., 2002; Koop, 1997; Koop, 1999; Korn & Scherner, 2000; Kowallik & Borbach-Jaene, 2001; Kruckenberg & Borbach-Jaene, 2001; Kruckenberg & Jaene, 1999; Leddy, Higgins & Naugle, 1999; Lekuona, 2001; Meek et al., 1993; Menzel, 2002; Menzel & Pohlmeier, 1999; Müller & Illner, 2002; Musters, Noordervliet & Keurs, 1996; Orloff & Flannery, 1996; Osborn et al., 1996; Pedersen & Poulsen, 1991a; Percival, 2000; Phillips, 1994; Reichenbach, 2002; Reichenbach, 2003; Reichenbach & Schadek, 2003; Reichenbach & Sinning, 2003; Sachslehner & Kollar, 1997; Scherner, 1999a; Schmidt et al., 2003; Schreiber, 1992; Schreiber, 1993a; Schreiber, 1993c; Schreiber, 1999; Schreiber, 2002; SEO, 1995; SGS Environment, 1994; Sinning, 1999; Sinning & Gerjets, 1999; Smallwood & Thelander, 2004; Sommerhage, 1997; Steiof, Becker & Rathgeber, 2002; Still, Little & Lawrence, 1996; Strickland et al., 2001b; Stübing & Bohle, 2001; Thelander & Rugge, 2000; Thelander, Smallwood & Rugge, 2003; Trapp et al., 2002; van der Winden, Spaans & Dirksen, 1999; Vierhaus, 2000; Walter & Brux, 1999; Winkelman, 1989; Winkelman, 1992a; Winkelman, 1992b; Young et al., 2003a; Young et al., 2003b

Because the intention was to produce a report primarily relevant to Germany, the main emphasis was on research from Germany. However, the number of studies included from other countries also reflects the scale of the research carried out in each country (Tab.1).

Table 1. Countries of the 127 studies evaluated in this report.

Country	Number of studies
Belgium	4
Germany	75
Denmark	2
France	2
Great Britain	6
Netherlands	5
Austria	2
Spain	10
USA	27
Australia	2

More than 2/3 of the analysed data was collected only during the operation of the wind farm, therefore no information about the situation before the installation of the wind farm is provided and consequently also no information about changes due to the construction. Results from separate control areas, which are used to distinguish influences of wind farms from more general factors (e.g. higher mortality caused by a cold winter), were only available in a small number of studies (Tab. 2). Nineteen of the 127 studies included the construction period of the wind farms.

Material and Methods

Number of studies	Phase		
8	pre-construction	during operation	and with separate control site
23	pre-construction	during operation	
9		during operation	and with separate control site
87		only during operation	

Table 2. Designs of studies (study in pre-construction period, control sites) on the impact of wind farms.

Nearly all studies refer to wind farms, only a few to single wind turbines. There was not enough material to distinguish between wind farms and single wind turbines.

The average duration of the 127 studies was 2.8 years, where one year counts as one season (breeding time, autumn migration etc.) and studies which continued into a new calendar year were counted for pragmatic reasons as two years. The range was 1 to 17 years. 51 studies, more than a third, only covered one season (Fig. 1).

Most of the studies apply to several bird or bat species. Often for each species several parameters (for example, minimal distance to wind farms and change in the resting population after the installation of a wind farm; for further details see below) were analysed. A comparison of species and parameters led to a data matrix with 1789 data sets. Only around a third are quantitative analyses, the remaining data sets are „single observations". Many of these „single observations" had their source in systematic surveys, but within the framework of which certain bird species were only rarely observed.

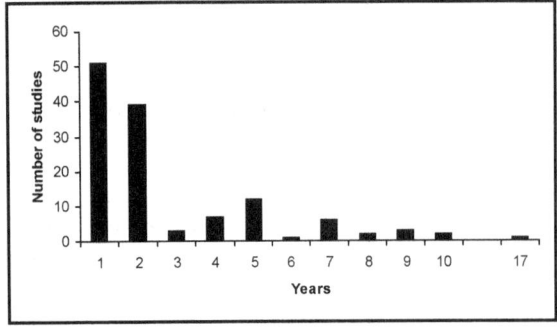

Figure 1. Number of study years of 127 wind farm impact studies

As expected, most of the data relate to birds of the open habitats (Tab. 3); woodland birds were rarely recorded. Even so, 39 species appear at least 10 times in the data tables.

The 127 studies analysed differ considerably both in their approach to and their extent of observation. The survey intensity ranged from casual observations to very well-founded, long-running (several years) projects. Overall, only a few of the studies are scientifically robust enough in their own right to withstand statistical tests for effects of wind farms on bird populations. The minimum requirements for such studies are at least two years (or two seasons) of pre-construction monitoring of the wind farm site and at least the same amount of monitoring once the wind farm is operating. In addition, contemporaneous parallel studies on control areas with no wind farms are necessary, so that further factors (for example land use) may be taken into account. These standards apply to the scientific evidence of impacts of a single wind turbine on bird populations. It should not be confused with investigations to prove whether a particular area might be suitable for wind energy development, as other criteria apply here.

References were only suitable for a formal meta-analysis (Fernandez-Duque & Valeggia, 1994), if the observation lasted at least four years, or if they use an appropriate number of control areas. Because this was only the case for a few studies, alternative methods needed to be used: (1) restricting the analysis to the studies with statistically significant results; or (2) evaluating data from all studies, irrespective of their quality and scope. For reasons given above, only a few studies were suitable for statistical analysis (alternative (1)), therefore, for further interpretation, alternative (2) has been chosen.

Material and Methods

In this procedure, all available results were included in the analysis, whether or not they were derived from systematic survey, or were based on only a few casual observations. The disadvantage of using all available studies is that both casual observations and extensive research are in statistical terms treated equally. It cannot be ruled out that „extreme" observations have been recorded more frequently than less spectacular events. Additional factors, which could be important in individual cases, also may not have been fully considered. However, the advantage of this method is that the number of studies included in the analysis is large and therefore the results are less dependent on single, well-researched, but possibly atypical studies. The independence of the data is also guaranteed. With a large number of studies the chances are improved that confounding factors cancel each other out.

Unless stated otherwise, the statistical tests in this report use the „null-hypothesis" that wind farms have no influence on the parameter under consideration (such as population size before and after the wind farm installation). The alternative hypothesis is that wind farms do have an influence. The tests determine how many stu-

Species		Number of records
Lapwing	*Vanellus vanellus*	80
Skylark	*Alauda arvensis*	55
Black-headed gull	*Larus ridibundus*	47
Starling	*Sturnus vulgaris*	40
Golden plover	*Pluvialis apricaria*	39
Mallard	*Anas platyrhynchos*	37
Buzzard	*Buteo buteo*	35
Carrion crow	*Corvus corone corone*	35
Curlew	*Numenius arquata*	33
Kestrel	*Falco tinnunculus*	32
Meadow pipit	*Anthus pratensis*	28
Herring gull	*Larus argentatus*	26
Oystercatcher	*Haematopus ostralegus*	25
Linnet	*Carduelis cannabina*	25
Blackbird	*Turdus merula*	22
Woodpigeon	*Columba palumbus*	22
Reed bunting	*Emberiza schoeniclus*	20
Common gull	*Larus canus*	18
White wagtail	*Motacilla alba*	15
Grey heron	*Ardea cinerea*	15
Chaffinch	*Fringilla coelebs*	14
Whitethroat	*Sylvia communis*	14
Yellowhammer	*Emberiza citrinella*	14
Red kite	*Milvus milvus*	14
Redshank	*Tringa totanus*	14
Yellow wagtail	*Motacilla flava*	13
Griffon vulture	*Gyps fulvus*	12
Cormorant	*Phalacrocorax carbo*	12
Marsh warbler	*Acrocephalus palustris*	12
Black-tailed godwit	*Limosa limosa*	12
Jackdaw	*Corvus monedula*	11
Wigeon	*Anas penelope*	11
Grey partridge	*Perdix perdix*	11
Tufted duck	*Aythya fuligula*	11
Reed warbler	*Acrocephalus scirpaceus*	11
Fieldfare	*Turdus pilaris*	11
Blue tit	*Parus caeruleus*	10
Whinchat	*Saxicola rubetra*	10
Swallow	*Hirundo rustica*	10

Table 3. Bird species with at least 10 records in the database.

dies had negative or positive effects (e.g. decrease or increase of the population). As mentioned above, neither the strength of effect, nor statistical significance have been considered. Neutral results (e.g. constant population) have been classified as positive, in order to avoid any false association of wind energy with negative impacts. At the same time, statistically significant negative effects shall be made more convincing and safer, and are not „diluted" by the inclusion of neutral results. If wind farms have no influence on bird populations, one would expect roughly equal proportions of positive and negative effects. If the frequency of positive and negative effects differs strongly, an impact of wind energy can be assumed. In these cases, the statistical test used is the binomial test (Sachs, 1978). Because not all of the available information is used in this procedure (for example the strength of the effects), it is very conservative, meaning that differences and trends are only classed as significant when they are very strong. The statistical tests were carried out using SPSS 7.5 statistical software.

Because the individual bird and bats species differ greatly in their biology and their use of habitat, the evaluation was carried out whenever possible for separate species. In cases when such differentiation was not possible species were grouped.

It is assumed that animals which are comparatively bounded to their breeding areas react differently to wind turbines from the ones passing through areas outside the breeding season, when they are less dependent on the resources of a single area and lack local knowledge. Therefore, it has been distinguished whether studies were carried out during or outside the breeding season (definition varies, depending which species is looked at). Most studies did not indicate which activities the animals were carrying out at the time of the observation (e.g. foraging, resting, roosting) and therefore, this factor could not be considered in this report.

The results have been divided into the following categories: „impact on bird population"; „disturbance effects"; „barrier effects on migrating and flying birds"; and „collision rates", because the studies analysed suggested such a division and because other parameters have been rarely studied.

The category „disturbance effects" analyses bird and bat activity on the ground and its minimum distance from wind farms; the category „barrier effects" refers both to migration and to regular flights of birds, for example between feeding and roosting places.

3 Impacts of wind farms on vertebrates

3.1 Non-lethal impacts (disturbance, displacement, habitat loss) on birds

3.1.1 Change in distributions due to wind farms

In order to test if wind farms have an impact on bird populations, only activity taking place on the ground or in the vegetation has been included. Migrating birds or birds on regular migration movements will be considered in section 3.1.3. Because only relatively few studies include a before-after-control impact comparison, studies comparing bird populations on the wind farm site with bird populations at similar sites in the surrounding area have also been used. The studies are not suitable each in their own right to prove effects of wind farms, because the habitat composition of control areas is never the same as the analysed areas. Even a before-after-comparison is not conclusive, because other factors, e.g. weather or supra-regional trends, could be responsible for changes in population size. As the analysed studies differ greatly from each other, for the purposes of interpreting the data only positive or negative effects of wind farms have been taken into account. Negative effects are (1) population decline after installation of the wind farm or (2) reduced numbers of birds within the wind farm or the surrounding area in comparison to control areas. Positive effects are accordingly population increase after the construction of the wind farm, or increased bird numbers around the wind farm. The strength of the effects has not been considered. If no population differences were detected, the effects were classified as positive in order to avoid falsely inflating the impact of negative effects (see above).

If there are no impacts of wind energy, equal ratios of positive and negative effects are expected. Statistical significance has been tested using a binomial test (for which the null hypothesis is that data are randomly distributed) (Table 4).

Data from 40 species which were included in at least six studies were good enough to be included in statistical tests (Tab. 4). Negative population impacts of wind farms during the breeding season could not be verified for any bird species. Only waders and gamebirds displayed reduced numbers in connection with wind farms. Positive or neutral effects predominate for the remaining species. Two species, which breed in reeds (marsh warbler and reed bunting) even showed significantly more positive or neutral reactions towards wind farms than negative reactions.

Studies carried out outside the breeding season show a very different picture. The negative impacts of wind farms predominate and geese (bean goose, white-fronted goose, greylag goose and barnacle goose), wigeons, lapwings and golden plovers display significantly more negative than positive effects. The main exception is starling, for which significantly more positive (or neutral) effects were recorded.

Overall, this study confirms statistically the results of others (Langston, 2002; Reichenbach, 2003), namely than wind farms have less impact on breeding birds, but more impacts on non-breeding birds.

3.1.2 Minimum avoidance distance of birds to wind farms

Many studies report the avoidance of wind farms by birds, as well as distances involved. For 28 species, with at least five studies each, the median and mean values of the minimum distances were calculated (Tab. 5). Some of the studies are the same as those used in the previous chapter („impacts of wind farms on populations").

The data show much variation, which may be seen by comparing results between species and „within" species. Therefore, in some cases standard deviations are very high (Tab. 5). Figures 2 to 13 confirm high variances for some results. This may be explained either by the use of casual observations, which naturally show higher dispersion, or by large differences between individual wind farms (see below).

Despite the high degree of variation, some trends are clearly noticeable. Avoidance distances during the breeding season were smal-

Table 4. Impacts of wind farms on bird populations as revealed from literature. The figures show the numbers of studies with positive or negative effects. Positive effect: 1) Density of birds higher or equal after construction of the wind farms or 2) density of birds close to wind farm higher or equal to density of control sites. Grey shadings indicate predominating negative effects. The last column gives the result of sign tests.

	Species	No negative effect	Negative effect	Significance
\multicolumn{5}{c}{Breeding season}				
Mallard	Anas platyrhynchos	6	5	ns
Common buzzard	Buteo buteo	3	3	ns
Grey partridge	Perdix perdix	4	5	ns
Quail	Coturnix coturnix	1	5	ns
Black-tailed godwit	Limosa limosa	5	6	ns
Redshank	Tringa totanus	2	9	ns
Oystercatcher	Haematopus ostralegus	6	7	ns
Lapwing	Vanellus vanellus	11	18	ns
Skylark	Alauda arvensis	15	15	ns
Meadow pipit	Anthus pratensis	15	7	ns
Yellow wagtail	Motacilla flava	7	3	ns
White wagtail	Motacilla alba	4	4	ns
Whinchat	Saxicola rubetra	2	6	ns
Stonechat	Saxicola torquata	5	1	ns
Blackbird	Turdus merula	5	4	ns
Wren	Troglodytes troglodytes	5	1	ns
Willow warbler	Phylloscopus trochilus	4	2	ns
Chiffchaff	Phylloscopus collybita	4	2	ns
Sedge warbler	Acrocephalus schoenobaenus	8	0	0.05
Reed warbler	Acrocephalus scirpaceus	6	1	ns
Marsh warbler	Acrocephalus palustris	6	4	ns
Whitethroat	Sylvia communis	8	4	ns
Blue tit	Parus caeruleus	4	3	ns
Yellowhammer	Emberiza citrinella	4	5	ns
Reed bunting	Emberiza schoeniclus	10	2	0.05
Linnet	Carduelis cannabina	2	6	ns
Carrion crow	Corvus corone	6	2	ns
Blackbird	Turdus merula	5	4	ns
\multicolumn{5}{c}{Non-breeding season}				
Grey heron	Ardea cinerea	5	1	ns
Wigeon	Anas penelope	0	9	0.01
Mallard	Anas platyrhynchos	3	7	ns
Tufted duck	Aythya fuligula	2	6	ns
Red kite	Milvus milvus	3	4	ns
Common buzzard	Buteo buteo	10	10	ns
Kestrel	Falco tinnunculus	13	7	ns
Curlew	Numenius arquata	11	19	ns
Oystercatcher	Haematopus ostralegus	4	3	ns
Lapwing	Vanellus vanellus	12	29	0.05
Golden plover	Pluvialis apricaria	8	21	0.05
Common gull	Larus canus	3	5	ns
Herring gull	Larus argentatus	2	4	ns
Black-headed gull	Larus ridibundus	14	5	ns
Woodpigeon	Columba palumbus	1	6	ns
Skylark	Alauda arvensis	4	2	ns
Fieldfare	Turdus pilaris	1	5	ns
Starling	Sturnus vulgaris	17	5	0.05
Jackdaw	Corvus corone	12	7	ns
Goose spp.		1	12	0.01

ler than outside the breeding season. Only a few wader species avoided close contact with wind farms at all times of year. The unusually high breeding season avoidance distance for black-tailed godwit may be due to chance, because this species is relatively rare. There is no direct evidence so far that the population of black-tailed godwit has been reduced by wind farms (Ketzenberg et al., 2002). The diagrams of the breeding skylark and reed bunting demonstrate that distances of more than 200m were exceptional; the majority of birds were present in the immediate vicinity of wind farms.

Higher avoidance distances from wind farms were generally observed outside the breeding season. As expected, birds of open habitats, e.g. geese, ducks and waders, generally avoided turbines by several hundred meters. Geese were particularly sensitive. Remarkable exceptions were grey heron, birds of prey, oystercatcher, gulls, starling and crows, which were frequently observed close to or within wind farms.

The sensitive species roosted at least 400-500m from wind farms (Tab. 5). Greater avoidance distances are likely to cause negative effects only in exceptional circumstances. To a large extent, the results correspond with the conclusions of single studies on this topic (Kruckenberg & Jaene, 1999; Reichenbach, 2003; Schreiber, 1993b; Schreiber, 1999).

When assessing the results, it should be remembered that it was only possible to analyse a large number of studies for a small number of species. Many species were hardly ever or even never examined. This is particularly so for the more controversial species (storks, birds of prey, cranes). Therefore, the list of species sensitive to disturbance is not complete.

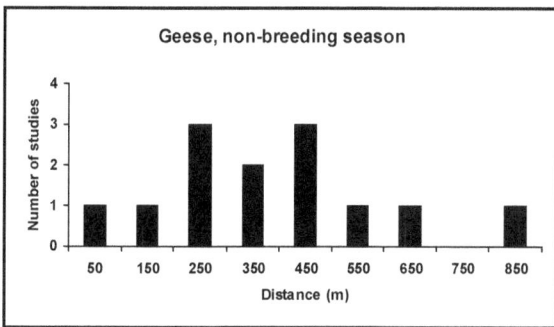

Figure 2. Minimal distances to wind farms of geese during the non-breeding season. The heights of the columns show the numbers of studies. The minimum distances to wind farms (or the distances up to which disturbances could be noticed) are shown on the x-axis.

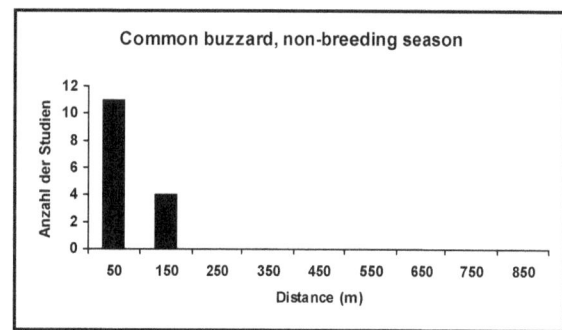

Figure 3. Minimal distances to wind farms of Common Buzzards during the non-breeding season. The heights of the columns show the numbers of studies. The minimum distances to wind farms (or the distances up to which disturbances could be noticed) are shown on the x-axis.

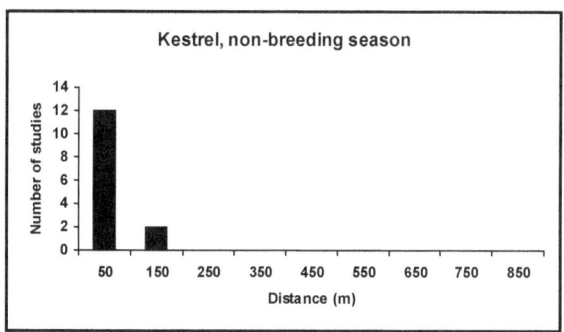

Figure 4. Minimal distances to wind farms of Kestrels during the non-breeding season. The heights of the columns show the numbers of studies. The minimum distances to wind farms (or the distances up to which disturbances could be noticed) are shown on the x-axis.

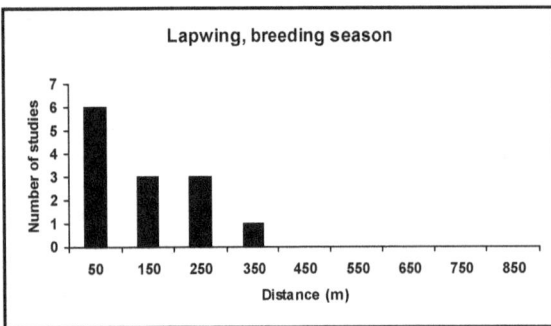

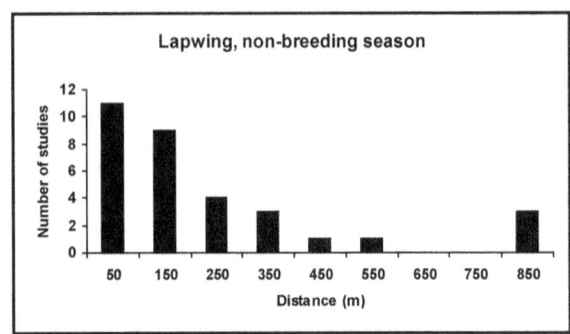

Figure 5. Minimal distances to wind farms of Lapwings during the breeding season. The heights of the columns show the numbers of studies. The minimum distances to wind farms (or the distances up to which disturbances could be noticed) are shown on the x-axis.

Figure 6. Minimal distances to wind farms of Lapwings during the non-breeding season. The heights of the columns show the numbers of studies. The minimum distances to wind farms (or the distances up to which disturbances could be noticed) are shown on the x-axis.

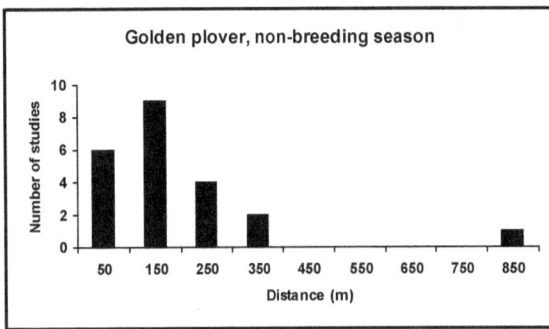

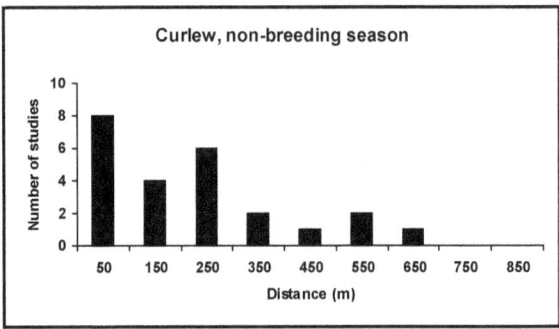

Figure 7. Minimal distances to wind farms of Golden Plovers during the non-breeding season. The heights of the columns show the numbers of studies. The minimum distances to wind farms (or the distances up to which disturbances could be noticed) are shown on the x-axis.

Figure 8. Minimal distances to wind farms of Curlews during the non-breeding season. The heights of the columns show the numbers of studies. The minimum distances to wind farms (or the distances up to which disturbances could be noticed) are shown on the x-axis.

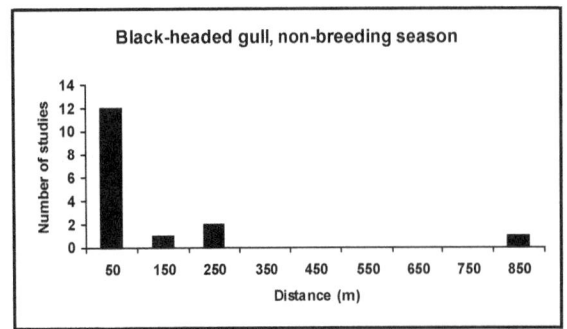

Figure 9. Minimal distances to wind farms of Black-headed Gulls during the non-breeding season. The heights of the columns show the numbers of studies. The minimum distances to wind farms (or the distances up to which disturbances could be noticed) are shown on the x-axis.

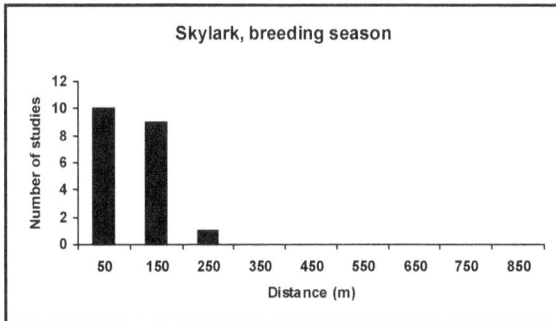

Figure 10. Minimal distances to wind farms of Skylarks during the breeding season. The heights of the columns show the numbers of studies. The minimum distances to wind farms (or the distances up to which disturbances could be noticed) are shown on the x-axis.

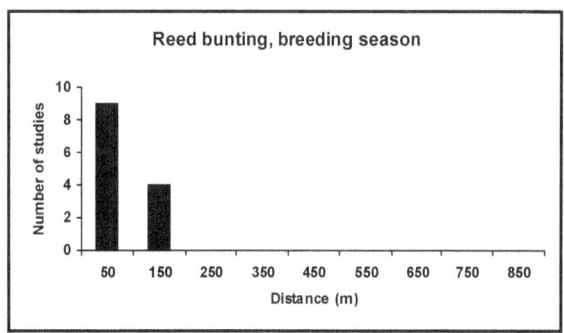

Figure 11. Minimal distances to wind farms of Reed Buntings during the breeding season. The heights of the columns show the numbers of studies. The minimum distances to wind farms (or the distances up to which disturbances could be noticed) are shown on the x-axis.

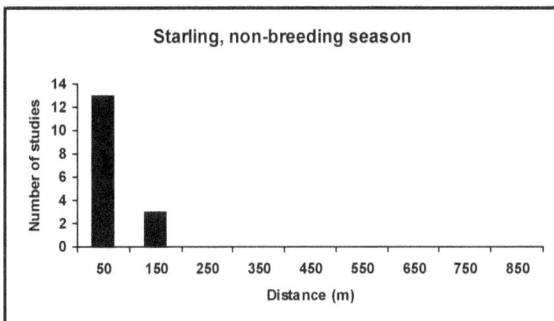

Figure 12. Minimal distances to wind farms of Starlings during the non-breeding season. The heights of the columns show the numbers of studies. The minimum distances to wind farms (or the distances up to which disturbances could be noticed) are shown on the x-axis.

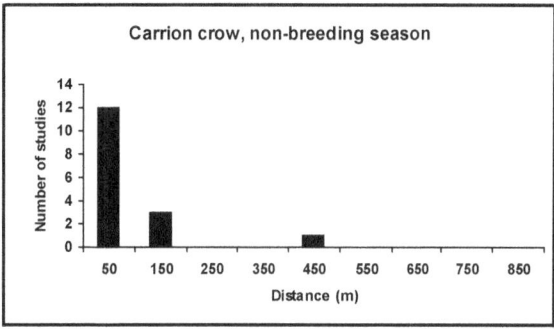

Figure 13. Minimal distances to wind farms of Carrion Crows during the non-breeding season. The heights of the columns show the numbers of studies. The minimum distances to wind farms (or the distances up to which disturbances could be noticed) are shown on the x-axis.

Habituation of birds to wind farms

Animals can become accustomed to certain types of repeated disturbance. In the case of wind farms this could mean that distances by which birds avoid turbines slowly reduce during the years after installation. The negative effects would get less and fewer birds would be displaced from their area. The term „habituation" is not used here in the strict behavioural sense, but rather colloquially. In ethology, habituation means the capacity of an animal to become accustomed to and no longer react towards repeated disturbance, related neither to positive nor to negative consequences (Immelmann, 1976). The term is used here to describe a [hypothetical] phenomenon whereby the distribution of birds becomes closer to wind farms

Table 5. Minimal distances to wind farms in studies of different bird species.

Species		Number of studies	Median	Mean	SD
Breeding season					
Mallard	*Anas platyrhynchos*	8	113	103	56
Black-tailed godwit	*Limosa limosa*	5	300	436	357
Oystercatcher	*Haematopus ostralegus*	8	25	85	113
Lapwing	*Vanellus vanellus*	13	100	108	110
Redshank	*Tringa totanus*	6	188	183	111
Skylark	*Alauda arvensis*	20	100	93	71
Meadow pipit	*Anthus pratensis*	9	0	41	53
Yellow wagtail	*Motacilla flava*	7	50	89	107
Blackbird	*Turdus merula*	5	100	82	76
Willow warbler	*Phylloscopus trochilus*	5	50	42	40
Chiffchaff	*Phylloscopus collybita*	5	50	42	40
Sedge warbler	*Acrocephalus schoenobaenus*	7	0	14	24
Reed warbler	*Acrocephalus scirpaceus*	11	25	56	70
Marsh warbler	*Acrocephalus palustris*	9	25	56	68
Whitethroat	*Sylvia communis*	9	100	79	65
Reed bunting	*Emberiza schoeniclus*	13	25	56	70
Linnet	*Carduelis cannabina*	5	125	135	29
Non-breeding season					
Grey heron	*Ardea cinerea*	6	30	65	97
Wigeon	*Anas penelope*	9	300	311	163
Swan spp.		8	125	150	139
Goose spp.		13	300	373	226
Mallard	*Anas platyrhynchos*	9	200	161	139
Diving ducks		12	213	219	122
Common buzzard	*Buteo buteo*	15	25	50	53
Kestrel	*Falco tinnunculus*	14	0	26	45
Curlew	*Numenius arquata*	24	190	212	176
Oystercatcher	*Haematopus ostralegus*	6	15	55	81
Lapwing	*Vanellus vanellus*	32	135	260	410
Common snipe	*Gallinago gallinago*	5	300	403	221
Golden plover	*Pluvialis apricaria*	22	135	175	167
Woodpigeon	*Columba palumbus*	5	100	160	195
Common gull	*Larus canus*	6	50	113	151
Black-headed gull	*Larus ridibundus*	15	0	97	211
Skylark	*Alauda arvensis*	6	0	38	59
Starling	*Sturnus vulgaris*	16	0	30	54
Carrion crow	*Corvus corone*	16	0	53	103

over time. There was no evidence to show if habituation is founded on an individual basis.

11 studies provided data with at least two years of observation after the installation of the wind farm. Each study analysed several species, which altogether resulted in 122 data sets. Only a few studies explicitly referred to „habituation". Even if observations suggested habituation (closer distribution in relation to wind farms after a few years; population increase in the area around wind farms some year after installation), it could not be completely ruled out that these were due to other factors, e.g. changes in habitat. Because of the variable quality of the data, the method, used elsewhere in this report has also been used here. The number of cases suggesting habituation (see above), have been counted and compared with the number of cases, which do not suggest habituation. If cases of habituation strongly predominate, one could talk of a widespread phenomenon. If this is not the

case, (e.g. the cases with or without habituation are balanced or those without habituation predominate), then it is questionable whether birds can get used to wind farms on a large scale.

For breeding birds, 38 out of 84 cases indicate habituation (45%, so less than half). For resting birds, the corresponding numbers are 25 out of 38 cases. More than half of the resting birds (66%) therefore seem to get used to wind farms. None of the results differ significantly from random, effectively a balance between cases with and without habituation.

Only in a few cases were sufficient data available for individual species. During the breeding season, results of six studies on lapwing indicate no habituation while two cases assume habituation. Outside the breeding season, three out of five studies suggest habituation. There were six studies for both skylark and meadow pipit; in each case, three studies indicated habituation.

The observed degree of habituation was in most cases very small. Even if it cannot ruled out that birds actually become habituated to wind farms, the data clearly indicate that this is neither a widespread nor a strong phenomenon.

Table 6. Number of studies with or without indications of habituation of birds to wind farms. A decreasing minimal distance between birds and wind farms in the course of study years is considered as an indication of habituation, the reverse is considered as the lack of habituation.

	Species	Number of cases without signs of habituation (increased distance to wind farm)	Number of cases with signs of habituation (decreased distance to wind farm)
Non-breeding season			
White-fronted goose	Anser albifrons	1	0
Wigeon	Anas penelope	0	1
Mallard	Anas platyrhynchos	0	1
Eider	Somateria mollissima	0	2
Common scoter	Melanitta nigra	0	2
Red kite	Milvus milvus	1	0
Common buzzard	Buteo buteo	1	1
Kestrel	Falco tinnunculus	1	1
Coot	Fulica atra	0	1
Oystercatcher	Haematopus ostralegus	0	1
Curlew	Numenius arquata	4	0
Lapwing	Vanellus vanellus	2	3
Golden plover	Pluvialis apricaria	1	3
Common gull	Larus canus	1	1
Herring gull	Larus argentatus	1	0
Black-headed gull	Larus ridibundus	1	1
Woodpigeon	Columba palumbus	1	1
Starling	Sturnus vulgaris	1	0
Carrion crow	Corvus corone	1	2

Species		Number of cases without signs of habituation (increased distance to wind farm)	Number of cases with signs of habituation (decreased distance to wind farm)
Breeding season			
Mallard	*Anas platyrhynchos*	0	2
Grey partridge	*Perdix perdix*	0	4
Quail	*Coturnix coturnix*	0	1
Pheasant	*Phasianus colchicus*	0	1
Moorhen	*Gallinula chloropus*	0	1
Oystercatcher	*Haematopus ostralegus*	2	2
Common snipe	*Gallinago gallinago*	0	1
Curlew	*Numenius arquata*	1	0
Black-tailed godwit	*Limosa limosa*	1	2
Redshank	*Tringa totanus*	3	1
Lapwing	*Vanellus vanellus*	6	2
Woodpigeon	*Columba palumbus*	1	0
Skylark	*Alauda arvensis*	3	3
Meadow pipit	*Anthus pratensis*	3	3
White wagtail	*Motacilla alba*	1	0
Yellow wagtail	*Motacilla flava*	0	2
Red-backed shrike	*Lanius collurio*	0	1
Dunnock	*Prunella modularis*	0	1
Wheatear	*Oenanthe oenanthe*	0	1
Whinchat	*Saxicola rubetra*	1	2
Stonechat	*Saxicola torquata*	1	1
Redstart	*Phoenicurus phoenicurus*	1	0
Spotted flycatcher	*Muscicapa striata*	1	0
Wren	*Troglodytes troglodytes*	0	1
Mistle thrush	*Turdus viscivorus*	1	0
Blackbird	*Turdus merula*	1	0
Song thrush	*Turdus philomelos*	1	0
Grasshopper warbler	*Locustella naevia*	1	1
Sedge warbler	*Acrocephalus schoenobaenus*	1	2
Reed warbler	*Acrocephalus scirpaceus*	0	2
Marsh warbler	*Acrocephalus palustris*	1	0
Icterine warbler	*Hippolais icterina*	1	0
Blackcap	*Sylvia atricapilla*	1	0
Garden warbler	*Sylvia borin*	1	0
Whitethroat	*Sylvia communis*	1	0
Chiffchaff	*Phylloscopus collybita*	1	0
Willow warbler	*Phylloscopus trochilus*	1	0
Great tit	*Parus major*	1	0
Blue tit	*Parus caeruleus*	1	0
Yellowhammer	*Emberiza citrinella*	1	0
Reed bunting	*Emberiza schoeniclus*	2	0
Greenfinch	*Carduelis chloris*	0	1
Chaffinch	*Fringilla coelebs*	1	0
Tree sparrow	*Passer montanus*	0	1
Starling	*Sturnus vulgaris*	1	0
Nutcracker	*Garrulus glandarius*	1	0

If the impacts of individual wind farms appear to increase over the years, this could be explained by the reluctance of breeding birds to give up their territory immediately after the installation of the wind farm. As territories with wind farms become less attractive to future generations in this area, in time this results in a thinning population density around the wind farms. It was not possible, on the basis of existing data, to determine if such a phenomenon actually occurs.

Avoidance distance and height of turbine

As mentioned above, impacts on bird populations differ between wind farms. The height of wind turbines appears to be at least partly responsible for these differences. The question of how the height of wind turbines relates to avoidance distances is also highly relevant to „repowering".

Turbine hub height is perhaps the most significant factor influencing birds. Hub height is closely correlated with the capacity (wattage) of the wind turbine. The studies analysed in this report show the following significant statistically relationship (regression function, Fig. 14):

Hub height (m) = 65.22 * power (MW)$^{0.457}$
$R^2 = 0.73$ (n = 78; p<0.001)

(R: correlation coefficient; n = sample size; p: error probability)

The relationship between hub height and avoidance distance has been calculated (Tab. 7) for bird species for which distance observations were available from at least four different wind farms (minimum number needed to reach a statistically significant result).

With the exception of the lapwing outside the breeding season, none of the results were statistically significant. Non-breeding lapwings are obviously very sensitive to very large wind turbines. The relationship between turbine height and avoidance distance was nearly linear.

Despite the fact that none of the remaining results presented in Tab. 7 were statistically significant, their overall tendency is still clear. Breeding birds, particularly songbirds, but also oystercatcher and redshank, are less affected by tall turbines than by small ones. Only lapwing and black-tailed godwit clearly avoid larger turbines.

Resting birds show a different picture. With a few exceptions (grey heron, diving ducks, oystercatcher and common snipe), the higher the turbines, the greater the avoidance distance. It may be that differences between breeding and non-breeding seasons are in fact the result of observing different suites of species at these different times of year. Breeding season observations were mainly of songbirds, whereas non-breeding observations tended to be of larger species of open habitats.

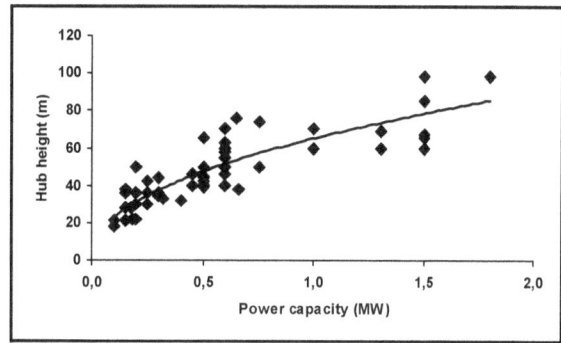

Figure 14. Relationship between tower height and power capacity of wind turbines.

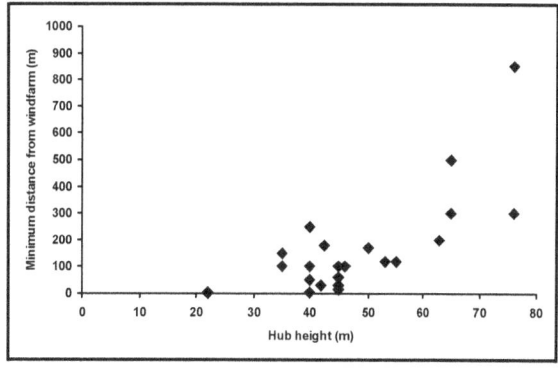

Figure 15. Relationship between minimal distances to wind farms of Lapwings and tower height. The relationship is statistically significant (n=24; R2=0.53; p<0.001).

Table 7. Relationships between minimal distances to wind farms of different bird species and tower height. The right-hand column indicates the change of minimal distance to wind farms when tower height is increased by one meter. Statistically significant relationships are marked by bold characters (only Lapwing during the non-breeding season).

	Species	n	Change of minimal distance to wind farms when tower height increased by 1 metre
	Breeding season		
Mallard	Anas platyrhynchos	7	0.09
Black-tailed godwit	Limosa limosa	5	3.67
Oystercatcher	Haematopus ostralegus	8	-2.64
Lapwing	Vanellus vanellus	12	1.78
Redshank	Tringa totanus	6	-2.64
Skylark	Alauda arvensis	19	-1.6
Meadow pipit	Anthus pratensis	9	-1.17
Yellow wagtail	Motacilla flava	6	-0.02
Blackbird	Turdus merula	4	-1.07
Willow warbler	Phylloscopus trochilus	5	-0.32
Chiffchaff	Phylloscopus collybita	5	-0.32
Sedge warbler	Acrocephalus schoenobaenus	6	-0.95
Reed warbler	Acrocephalus scirpaceus	9	-0.51
Marsh warbler	Acrocephalus palustris	6	-1.67
Whitethroat	Sylvia communis	8	-1.47
Reed bunting	Emberiza schoeniclus	12	-3.41
Linnet	Carduelis cannabina	4	0.66
	Non-breeding season		
Grey heron	Ardea cinerea	6	-1.64
Wigeon	Anas penelope	7	0.41
Goose spp.	Geese	6	6.22
Mallard	Anas platyrhynchos	7	0.95
Diving ducks		10	-1.64
Common buzzard	Buteo buteo	12	1.29
Kestrel	Falco tinnunculus	10	0.88
Curlew	Numenius arquata	19	1.95
Oystercatcher	Haematopus ostralegus	6	-2.79
Lapwing	**Vanellus vanellus**	**25**	**9.59**
Common snipe	Gallinago gallinago	5	-4.55
Golden plover	Pluvialis apricaria	15	3.12
Woodpigeon	Columba palumbus	4	1.2
Black-headed gull	Larus ridibundus	12	1.33
Starling	Sturnus vulgaris	9	1.54
Carrion crow	Corvus corone	12	1.61

The reasons for the different impacts of large wind turbines can only be guessed at. Birds of open habitats appear to be more threatened by large objects than by smaller ones. In this context, blade flicker could also play a role. Songbirds appear less affected by large turbines, the rotors of which tend to sweep higher above the ground than those of small turbines. The rotor movements of large turbines are unlikely to affect the airspace close to ground level, which is used by many small birds. Another factor could be that habitat changes are more likely to have taken place under large turbines than under smaller ones. Often, the areas beneath wind turbines are left fallow, so that after some time herbaceous plants and bushes grow, and these may subsequently be used by small birds.

3.1.3 Barrier effects of wind farms on birds

Another acknowledged potential impact of wind farms on birds is that they may act as barriers to migrating birds or to birds commuting between different sites (breeding, feeding and resting areas). Publications and reports on this topic are summarised in Table 8. Again, single observations and extensive investigations have been combined. A barrier effect was assumed in quantitative studies if at least 5 % of the individuals or flocks showed a measurable reaction to wind farms. Because of variations in the methods used by different studies, it was not possible to describe the type of reaction by birds to wind farms. In all cases, reactions included observed alterations in flight direction or height, so that birds flew around or above wind farms. In some cases it was also observed that birds turned around, or that the flight formation broke up when confronted by a wind farm.

Table 8 summarises 168 daytime observations. Not enough data were available for the night-time, when much bird migration takes place. A barrier effect was determined in 104 out of 168 cases. Because it cannot safely be assumed that reaction and lack of reaction by birds to wind farms are reported equally, this ratio was not of great significance. A barrier effect could be determined for 81 species, a clear majority of those analysed. It is therefore a relatively common phenomenon, but does not manifest itself in the same way in all species. Geese, kites, cranes and many small bird species were particularly sensitive. Some large birds (cormorant, grey heron), ducks, some birds of prey (sparrowhawk, common buzzard, kestrel) gulls and terns, starling and crows were all less sensitive or less willing to change their original migration direction when approaching wind farms. These species or species groups also avoided wind farms less often (Tab. 5) and their local populations were less influenced by wind farms (Tab. 4).

Avoidance of wind farms means a higher energetic output for birds on migration or involved in regular daily flight movements. How important this is depends on how often these situations occur. In an extreme case, the wind farm could be located between resting, roosting and/or breeding areas, leading to a dislocation of a species' essential biotope (Isselbächer & Isselbächer, 2001; Steiof et al., 2002). Research on barrier effects of wind farms is inadequate. It could not be tested whether birds react to wind farms during the night, or when the rotors are at a standstill.

3.2 Non-lethal impacts (disturbance, displacement, habitat loss) on mammals

The impact of wind farms on population size and distribution of mammals has so far been very little studied. The results were not consistent.

For two bat species (serotine and noctule bats) a decline in activity after the construction of a wind farms was identified, while the common pipistrelle increased its activity (Bach, 2002). In one study, deer and hares showed less activity for sample areas with wind farms than for ones with none (Bergen, 2002a; Menzel & Pohlmeier, 1999), however, the results were not significant. In one study in the USA, the po-

Table 8 (next page). Number of studies showing whether wind farms are a barrier to bird migration or regular flights. The last column shows the results of sign tests (null-hypothesis: equal frequency of impacts and non-impacts).

Species		Barrier effect		Significance
		Yes	No	
Cormorant	Phalacrocarax carbo	2	4	ns
Grey heron	Ardea cinerea	4	3	ns
Black stork	Ciconia nigra	1	1	
White stork	Ciconia ciconia	2	1	
Bean goose	Anser fabalis	1	0	
White-fronted goose	Anser albifrons	3	0	
Greylag goose	Anser anser	2	0	
Barnacle goose	Branta leucopsis	1	0	
Geese	sum	7	0	0.05
Wigeon	Anas penelope	1	0	
Teal	Anas crecca	0	1	
Shoveler	Anas clypeata	0	1	
Mallard	Anas platyrhynchos	3	2	
Pochard	Aythya ferina	1	0	
Tufted duck	Aythya fuligula	1	0	
Eider	Somateria mollissima	1	1	
Ducks	sum	7	5	ns
Griffon vulture	Gyps fulvus	1	0	
Red kite	Milvus milvus	3	0	
Black kite	Milvus migrans	4	0	
Honey buzzard	Pernis apivorus	1	0	
Goshawk	Accipiter gentilis	1	1	
Sparrowhawk	Accipiter nisus	1	3	
Common buzzard	Buteo buteo	2	4	ns
Short-toed eagle	Circaetus gallicus	1	1	
Marsh harrier	Circus aeruginosus	4	0	
Hen harrier	Circus cyaneus	1	0	
Peregrine	Falco peregrinus	1	0	
Merlin	Falco columbarius	1	0	
Hobby	Falco subbuteo	1	0	
Kestrel	Falco tinnunculus	3	2	
Birds of prey	sum	25	11	0.05
Crane	Grus grus	5	0	
Purple sandpiper	Calidris maritima	0	1	
Calidris spp.	Calidris.spec.	0	1	
Common snipe	Gallinago gallinago	1	0	
Curlew	Numenius arquata	1	0	
Lapwing	Vanellus vanellus	5	1	ns
Golden plover	Pluvialis apricaria	2	1	
Waders	sum	10	3	ns
Common gull	Larus canus	2	2	
Lesser black-backed gull	Larus fuscus	0	3	
Herring gull	Larus argentatus	3	3	ns
Great black-backed gull	Larus marinus	0	1	
Black-headed gull	Larus ridibundus	3	5	ns
Black tern	Chlidonias niger	0	1	
Sandwich tern	Sterna sandvicensis	1	0	
Common tern	Sterna hirundo	3	1	
Little tern	Sterna albifrons	0	1	
Gulls & Terns	sum	12	17	ns

Species		Barrier effect		
		Yes	No	Significance
Collared dove	Streptopelia decaocto	1	0	
Rock/feral dove	Columba livia	0	1	
Stock dove	Columba oenas	2	0	
Woodpigeon	Columba palumbus	3	2	
Pigeons	sum	6	3	ns
Swift	Apus apus	2	0	
Bee-eater	Merops apiaster	1	0	
Great spotted woodpecker	Dendrocopos major	1	0	
Barn swallow	Hirundo rustica	4	0	
House martin	Delichon urbica	2	0	
Calandra lark	Melanocorypha calandra	1	0	
Woodlark	Lullula arborea	2	0	
Skylark	Alauda arvensis	5	1	ns
Meadow pipit	Anthus pratensis	2	1	
Red-throated pipit	Anthus cervinus	1	0	
Pipit spp.	Anthus spec.	1	0	
Grey wagtail	Motacilla cinerea	1	0	
White wagtail	Motacilla alba	3	0	
Wagtail spp.	Motacilla spec.	1	0	
		1	0	
Great grey shrike	Lanius excubitor	1	0	
Dunnock	Prunella modularis	2	0	
Black redstart	Phoenicurus ochruros	1	0	
Wheatear	Oenanthe oenanthe	1	0	
Mistle thrush	Turdus viscivorus	3	0	
Fieldfare	Turdus pilaris	4	1	
Blackbird	Turdus merula	2	1	
Ring ouzel	Turdus torquatus	2	0	
Redwing	Turdus iliacus	2	1	
Song thrush	Turdus philomelos	2	0	
Thrush spp.	Turdus spec.	2	0	
Bearded tit	Panurus biarmicus	0	1	
Long-tailed tit	Aegithalos caudatus	1	0	
Goldcrest	Regulus regulus	0	1	
Blue tit	Parus caeruleus	1	0	
Great tit	Parus major	1	0	
Yellowhammer	Emberiza citrinella	2	0	
Reed bunting	Emberiza schoeniclus	2	0	
Unidentified finches & buntings		1	1	
Serin	Serinus serinus	2	0	
Linnet	Carduelis cannabina	3	0	
Siskin	Carduelis spinus	2	0	
Goldfinch	Carduelis carduelis	3	0	
Greenfinch	Carduelis chloris	2	0	
Hawfinch	Coccothraustes. coccothraustes	1	0	
Chaffinch	Fringilla coelebs	3	0	
Brambling	Fringilla montifringilla	2	0	
unidentified finch	Carduelis spec.	1	1	
Tree sparrow	Passer montanus	1	0	
Passerines (excl. starling and crows)	sum	74	9	0.001
Starling	Sturnus vulgaris	3	3	ns
Nutcracker	Nucifraga caryocatactes	1	0	
Rook	Corvus frugilegus	2	0	
Jackdaw	Corvus monedula	2	1	
Carrion crow	Corvus corone	1	3	
Crows	sum	6	4	ns

pulations of some small mammal species (prairie dog, cottontail rabbit and prairie hare) were apparently encouraged by wind farm developments, presumably through indirect effects such as habitat change during the construction process. The populations of other species (pronghorn and ground-squirrel) did not change (Johnson et al., 2000).

3.3 Collision of birds and bats with wind farms
3.3.1 Collisions of birds with wind farms

The majority of extensive studies on collisions of birds with wind farms have been carried out in the USA. In Europe, this topic has been less comprehensively investigated. Because the studies have been carried out systematically and over a long period of time, it was possible to calculate collision rates (individual birds per turbine per year); which are summarised in Table 9. Most of the data were collected while the wind farms were operating normally. Data from shut down wind farms were not represented to the same extent. Not all studies used the same method. In particular, they differed in whether they have followed the investigation protocol of Anderson et al. (1999) and Morrison, (2002), which is now standard practice in the USA. Among other things, the protocol allows for search efficiency of fieldworkers and the likelihood of carcasses disappearing from the study area (for example by being scavenged) before being recorded. If search efficiency and premature disappearance of carcasses are not taken into account, the collision rate may be underestimated. Therefore, the data in Table 9 might tend to underestimate, rather than overestimate actual collision rates.

Several studies each were available for a number of wind farms, mainly from the USA. The present data sets partly overlap with each other. To guarantee the independence of the data sets, each wind farm was included only once in the statistical analysis, using either the most recent reports or those including the most extensive observations.

The collision rates varied greatly between different wind farms. For many wind farms no collisions or nearly none occurred. At other wind farms, collisions occurred with a frequency of more than 30 per year per turbine.

Mass collisions, similar to those known from lighthouses or other buildings (Crawford & Engholm, 2001; Erickson et al., 2002; Manville, 201; Ugoretz, 2001) could not be identified for individual turbines within wind farms. In Sweden, a maximum of 43 birds (migrating songbirds) were found in one night at an illuminated, but non-operational wind farm (Karlsson, 1983). In the USA, the maximum collision rate is 14 birds per turbine per night – also migrating songbirds (Eriksson et al., 2001). Because some wind farms contain a large number of turbines (over 5000 at the Altamont Pass in California), even relatively low collision rates [per turbine] result in numerically high overall losses. Barely half of the studies reveal collision rates of one bird per turbine per year; the median was 1.7 and the mean was 8.1 victims per turbine per year. Median and mean for birds of prey were 0.3 and 0.6 victims per turbine per year, respectively.

In addressing the question of what factors might have caused the very different collision rates, first of all the height of the individual wind farms should be considered. A weak, statistically insignificant relationship between hub height and collision rate was detected ($y=0.29x$; $R^2=0.08$; Figure 16). The position of the wind farm may be more influential. Two main points stand out. Wind farms on bare mountain ridges or where there is a sharp change in relief (for example at plateau edges), which are quite common in the USA and Spain, caused high casualty rates, in particular for birds of prey. In central Europe, wetlands stood out as having particularly high casualty rates. Collision rates of more than two birds per wind turbine per year were only recorded at wetlands or on mountain ridges. The influence of habitat (categorised as: wetlands; mountain ridges; or other) on the collision rates was statistically significant (Kruskal-Wallis Test; $Chi^2=7.27$; $df=2$; $p<0.05$).

The most comprehensive data on collision victims at wind farms has been collated by Tobias Dürr (Dürr, 2001; Dürr, 2004). Table 10 is based on his data, which includes unpublished records, and to which some of the latest literature has been added. No conclusions about rates of collision can be drawn from Table 10, because not only have the data been collected over different time periods, but also search effort differs greatly between regions/countries

Table 9. Collision rates of all birds and raptors (annual number of victims per turbine) in different wind farms.

Country	Windfarm	Habitat	Birds	Birds of prey	Remarks	Reference
Belgium	Oostdam te Zeebrugge	Wetland	24		Further studies in other years	Everaert Devos & Kuijken, 2003
Belgium	Boudewijnkanaal te Brugge	Wetland	35		Further studies in other years	Everaert et al., 2003
Belgium	Elektriciteitscentrale te Schelle	Wetland	18		Further studies in other years	Everaert et al., 2003
Denmark	Tjaereborg	Wetland	3			Pedersen & Poulsen, 1991b
Germany	Bremerhaven-Fischereihafen	Wetland	9			Scherner, 1999b
Netherlands	Kreekraak sluice	Wetland	3.7			Musters et al., 1996
Netherlands	Oosterbierum	Grassland	1.8			Winkelman, 1992a
Netherlands	Urk	Coast	1.7			Winkelman, 1989
Sweden	Näsudden	Grassland	0.7			Percival, 2000
Spain	PESUR Parque Eólico del Sur and Parque and Parque Eólico de Levantera	Mountain ridges	0.36	0.36		Barrios & Rodriguez, 2004 SEO 1995
Spain	E3 Energia Eólica del Estrecho	Mountain ridges	0.03	0.03		Barrios & Rodriguez, 2004 SEO 1995
Spain	Salajones	Mountain ridges	21.69	8.33		Lekuona, 2001
Spain	Izco-Albar	Mountain ridges	22.63	0.93		Lekuona, 2001
Spain	Alaiz-Echague	Mountain ridges	3.56	0.62		Lekuona, 2001
Spain	Guennda	Mountain ridges	8.47	0.2		Lekuona, 2001
Spain	El Perdón	Mountain ridges	64.26	0.36		Lekuona, 2001
Spain	Tarifa			0.03	0.03	Janss, 2000
UK	Bryn Tytli	Moorland, Grassland	0			Phillips, 1994
UK	Burgar Hill Orkney	Moorland, Grassland	0.15			Percival, 2000
UK	Haverigg Cumbria	Moorland, Grassland	0	0		Percival, 2000
UK	Blyth	Wetland	1	34		Still et al., 1996
UK	Ovenden Moor	Moorland, Grassland	0.04	0		Percival, 2000
UK	Cemmaes	Moorland, Grassland	0.04	0		Percival, 2000
USA	Buffalo Ridge	Grassland	0.98	0.012	Further studies in other years	Erickson et al., 2001
USA	Foote Creek Rim	Prairie	1.75	0.036	Further studies in other years	Erickson et al., 2001
USA	Vansycle	Farmland, Grassland	0.63	0	Further studies in other years	Erickson et al., 2001
USA	Altamont	Mountain ridges	0.87	0.24	Further studies in other years	Smallwood & Thelander, 2004
USA	Nine Canyon Wind Project	Prairie	3.59		Further studies in other years	Erickson et al., 2003
USA	Green Mt Searsburg	Mountain ridges	0	0		Erickson et al., 2001
USA	IDWGP Algona	Mountain ridges	0	0		Erickson et al., 2001
USA	Somerset County	Mountain ridges	0	0		Erickson et al., 2001
USA	San Gorgino	Mountain ridges	2	31		Erickson et al., 2001
USA	Solano County	Mountain ridges	54			Erickson et al., 2001
Australia	Tasmania	Coast	1.86	0		Hydro Tasmania

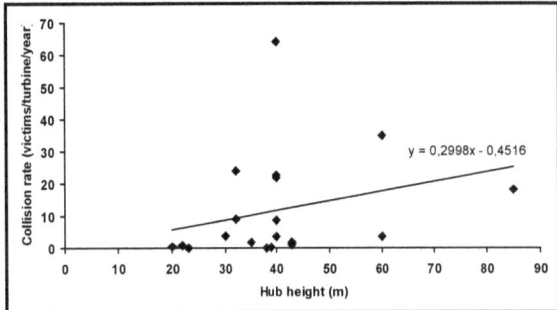

Figure 16. Collision rates of birds at wind farms of different tower heights.

The species composition of collision victims naturally depends on the make-up of the species assemblage of the area affected by wind farms. In the USA birds of prey, particularly golden eagle and red-tailed hawk, predominate at wind farms on the mountain ridges. This is also true at Spanish wind farms, where large numbers of griffon vultures were killed. In central and northern Europe different species were affected. Particularly striking was the large number of dead gulls, which are the fundamental reason for high collision rates near wetlands (Tab. 9). As well as gulls, birds of prey were strongly affected, particularly red kite, but also kestrel and common buzzard. Interestingly, red kites were killed mainly during the breeding season, while no losses were recorded of birds wintering in Spain (Tab. 10b). Also noteworthy is the number of 13 white-tailed eagles killed to date. Apart from the birds of prey, the starling stands out with a relatively high number of recorded casualties.

By comparing the number of casualties for individual species with their reactions towards wind farms (Tab. 4, 5 and 8), it is noticeable that species or species groups which are less afraid of wind farms are more likely to be collision victims than species which avoid or fly around wind farms by a wide margin. So, birds of prey, gulls and starlings were killed relatively frequently, while by comparison geese and waders were found less frequently as collision victims. An exception seems to be the crows, which are not afraid of wind farms, but also rarely get killed.

Table 10a. Number of bird collision victims found at wind farms in Germany since 1989 (intensified search since 2002). Data from Staatliche Vogelschutzwarte, LUA Brandenburg, T. Dürr, 1.11.2004. BB = Brandenburg; ST = Sachsen-Anhalt; SN = Sachsen; TH = Thüringen; MVP = Mecklenburg-Vorpommern; SH = Schleswig-Holstein (and Hamburg); NDS = Niedersachsen; HB = Bremen; NRW = Nordrhein-Westfalen; RP = Rheinland-Pfalz; HS = Hessen; SL = Saarland; BW = Baden-Württemberg; BY = Bayern.

Species		BB	ST	SN	TH	MVP	SH	NDS	HB	NRW	RP	HS	SL	BW	BY	Tot.
Red-throated diver	*Gavia stellata*							1								1
Cormorant	*Phalacrocorax carbo*							2								2
White stork	*Ciconia ciconia*	3				2	1							1		7
Black stork	*Ciconia nigra*											1				1
Whooper swan	*Cygnus cygnus*						1									1
Mute swan	*Cygnus olor*	1	1				1	5								8
Greylag goose	*Anser anser*							1								1
Bean goose	*Anser fabalis*			1												1
Bean/white-fronted goose	*Anser fabalis/albifrons*		1													1
Barnacle goose	*Branta leucopsis*						6									6
Shelduck	*Tadorna tadorna*							1								1
Mallard	*Anas platyrhynchos*			1			3	1	2							7
Teal	*Anas crecca*							1								1
Tufted duck	*Aythya fuligula*						1									1
White-tailed eagle	*Haliaeetus albicilla*	2	1			4	6									13
Red kite	*Milvus milvus*	20	10	4	1	1		1		1		3				41
Black kite	*Milvus migrans*	6														6
Goshawk	*Accipiter gentiles*	1														1
Sparrowhawk	*Accipiter nisus*	1														1
Common buzzard	*Buteo buteo*	15	5	2		1	2			1		1				27
Marsh harrier	*Circus aeruginosus*	1														1
Montagu's harrier	*Circus pygargus*											1				1
Kestrel	*Falco tinnunculus*	5	4	1												10

Species		BB	ST	SN	TH	MVP	SH	NDS	HB	NRW	RP	HS	SL	BW	BY	Tot.
Merlin	Falco columbarius	1														1
Hobby	Falco subbuteo/	1														1
Unidentified bird of prey		1														1
Grey partridge	Perdix perdix	1														1
Pheasant	Phasianus colchicus							1	1							2
Oystercatcher	Haematopus ostralegus						2	1								3
Golden plover	Pluvialis apricaria		2													2
Black-headed gull	Larus ridibundus	4					2	1	2							9
Common gull	Larus canus	2					1	2	2							7
Herring gull	Larus argentatus						9	2	1							12
Lesser black-backed gull	Larus fuscus						1									1
Guillemot	Uria aalge								1							1
Eagle owl	Bubo bubo									3				1		4
Woodpigeon	Columba palumbus	3	1													4
Rock/feral dove	Columba livia	6														6
Cuckoo	Cuculus canorus	1														1
Swift	Apus apus	6	2							1						9
Great spotted woodpecker	Dendrocopos major	1														1
Green woodpecker	Picus viridis	1														1
Skylark	Alauda arvensis	6														6
Barn swallow	Hirundo rustica	1														1
House martin	Delichon urbica	2														2
White wagtail	Motacilla alba	1														1
Yellow wagtail	Motacilla flava	1														1
Red-backed shrike	Lanius collurio	1														1
Wren	Troglodytes troglod.	1														1
Marsh warbler	Acrocephalus palustris							1								1
Robin	Erithacus rubecula	1														1
Pied flycatcher	Ficedula hypoleuca	2														2
Whinchat	Saxicola rubetra	1														1
Redwing	Turdus iliacus					1										1
Fieldfare	Turdus pilaris		1													1
Goldcrest	Regulus regulus									1						1
Firecrest	Regulus ignicapillus	1														1
Great tit	Parus major	1														1
Corn bunting	Emberiza calandra	9														9
Yellowhammer	Emberiza citrinella	3		1												4
Tree sparrow	Passer montanus	1														1
House sparrow	Passer domesticus	1														1
Greenfinch	Carduelis chloris	2														2
Starling	Sturnus vulgaris	4		1			1									6
Magpie	Pica pica		1													1
Raven	Corvus corax	9														9
Rook	Corvus frugilegus		1													1
Carrion crow	Corvus corone	2											1			3
Crow spp.	Corvus								1							1
Total		134	30	10	1	7	34	22	14	9	1	6	0	2	0	269

Table 10b. Number of bird collision victims found at wind farms in Europe. Data from Staatliche Vogelschutzwarte, LUA Brandenburg, T. Dürr, 06.09.2004 and from literature. NL: Netherlands; BE: Belgium; SEP: Spain; SWE: Sweden; AT: Austria; UK: Great Britain; DK: Denmark; D: Germany (as of July 2004).

Species		NL	BE	ESP	SWE	AT	UK	DK	D	Tot.
Red-throated diver	*Gavia stellata*								1	1
Cormorant	*Phalacrocorax carbo*								2	2
Grey heron	*Ardea cinerea*	2	1							3
White stork	*Ciconia ciconia*								6	6
Black stork	*Ciconia nigra*								1	1
Whooper swan	*Cygnus cygnus*								1	1
Mute swan	*Cygnus olor*				1				7	8
Domestic goose	*Anser a. domestica*		1							1
Greylag goose	*Anser anser*								1	1
Bean goose	*Anser fabalis*								1	1
Bean/white-fronted goose	*Anser fabalis/albifrons*								1	1
Barnacle goose	*Branta leucopsis*								6	6
Shelduck	*Tadorna tadorna*	1							1	2
Mallard	*Anas platyrhynchos*			11					7	18
Teal	*Anas crecca*	1							1	2
Tufted duck	*Aythya fuligula*								1	1
Duck spp.	*Anas sp*	1								1
Griffon vulture	*Gyps fulvus*			133						133
Booted eagle	*Hieraaetus pennatus*			1						1
Golden eagle	*Aquila chrysaetos*			1						1
White-tailed eagle	*Haliaeetus albicilla*								13	13
Short-toed eagle	*Circaetus gallicus*			2						2
Red kite	*Milvus milvus*				1		2		40	43
Black kite	*Milvus migrans*			1					6	7
Goshawk	*Accipiter gentiles*								1	1
Sparrowhawk	*Accipiter nisus*			1	1					2
Common buzzard	*Buteo buteo*				3				24	27
Marsh harrier	*Circus aeruginosus*								1	1
Montagu's harrier	*Circus pygargus*								1	1
Peregrine	*Falco peregrinus*		2							2
Hobby	*Falco columbarius*								1	1
Kestrel	*Falco tinnunculus*	4	2	13					10	29
Lesser kestrel	*Falco naumanni*				3					3
Merlin	*Falco columbarius*								1	1
Unidentified birds of prey					1				1	2
Red-legged partridge	*Alectoris rufa*			1						1
Grey partridge	*Perdix perdix*								1	1
Pheasant	*Phasianus colchicus*			3	1				2	6
Black grouse	*Tetrao tetrix*					2				2
Moorhen	*Gallinula chloropus*	1								1
Coot	*Fulica atra*	1	7							8
Oystercatcher	*Haematopus ostralegus*	4							3	7
Golden plover	*Pluvialis apricaria*	1			1				2	4
Lapwing	*Vanellus vanellus*	2								2
Redshank	*Tringa totanus*			1						1
Common snipe	*Gallinago gallinago*	1								1
Woodcock	*Scolopax rusticola*					1				1
Black-headed gull	*Larus ridibundus*	22	56						9	87
Kittiwake	*Rissa tridactyla*			1						1
Common gull	*Larus canus*	1	3		2			1	7	14
Herring gull	*Larus argentatus*	4	172		2				11	189

	Species	NL	BE	ESP	SWE	AT	UK	DK	D	Tot.
Lesser black-backed gull	*Larus fuscus*		44						1	45
Great black-backed gull	*Larus marinus*		6							7
Gull sp.	*Larus sp*	2			2				1	5
Common tern	*Sterna hirundo*		8							8
Little tern	*Sterna albifrons*		4							4
Guillemot	*Uria aalge*								1	1
Eagle owl	*Bubo bubo*			3					4	7
Woodpigeon	*Columba palumbus*	1	5	1	1				4	12
Stock dove	*Columba oenas*		1							1
Rock/feral dove	*Columba livia*		9						4	13
Pigeon sp.		1		2						3
Swift	*Apus apus*		2	1	3				8	14
Cuckoo	*Cuculus canorus*			1						1
Great spotted woodpecker	*Dendrocopos major*								1	1
House martin	*Delichon urbica*			1	6				1	8
Barn swallow	*Hirundo rustica*			1	1					2
White wagtail	*Motacilla alba*	1	1						1	3
Yellow wagtail	*Motacilla flava*								1	1
Woodlark	*Lullula arborea*			5						5
Crested lark	*Galerida cristata*			1						1
Skylark	*Alauda arvensis*			2					6	8
Tawny pipit	*Anthus campestris*			2						2
Robin	*Erithacus rubecula*	1	1	5					1	8
Marsh warbler	*Acrocephalus palustris*								1	1
Willow warbler	*Phylloscopus trochilus*				1				1	2
Pied flycatcher	*Ficedula hypoleuca*								2	2
Black redstart	*Phoenicurus ochrorus*			2						2
Whinchat	*Saxicola rubetra*								1	1
Stonechat	*Saxicola torquata*			1						1
Redwing	*Turdus iliacus*							1	1	2
Blackbird	*Turdus merula*		1	3	4				1	9
Song thrush	*Turdus philomelos*	1	4		1					6
Fieldfare	*Turdus pilaris*	1							1	2
Unidentified thrush	*Turdus*	1								0 [1]
Goldcrest	*Regulus regulus*			1					1	2
Firecrest	*Regulus ignicapillus*			1					1	2
Regulus sp.	*Regulus*	3								3
Whitethroat	*Sylvia communis*			1						1
Blackcap	*Sylvia atricapilla*			4						4
Great tit	*Parus major*								1	1
Magpie	*Pica pica*		1						1	2
Jackdaw	*Corvus monedula*	1								1
Raven	*Corvus corax*								9	9
Rook	*Corvus frugilegus*				1				1	2
Carrion crow	*Corvus corone*		1		1				3	5
Crow sp.	*Corvus sp*								1	1
Starling	*Sturnus vulgaris*	14	9						5	28
Corn bunting	*Emberiza calandra*								9	9
Yellowhammer	*Emberiza citrinella*								1	1
Tree sparrow	*Passer montanus*								1	1
House sparrow	*Passer domesticus*	3							1	4
Greenfinch	*Carduelis chloris*								2	2
Goldfinch	*Carduelis carduelis*	1								1
Chaffinch	*Fringilla coelebs*			1	1			1		3
Linnet	*Carduelis cannabina*			3				1		4
Common crossbill	*Loxia curvirostra*			1						1
Unidentified birds	*Aves sp*			4						4
Total		77	359	204	33	2	2	4	248	829

Mortality rates of birds

Only a few studies describe the extent to which collisions at wind farms increase annual mortality rates of the populations affected. Still et al. (1996) assumed that wind farms caused an additional mortality of 0.5-1.5% of the local eider population. Winkelmann (1992) estimated that the risk of a bird being killed while flying through a wind farm is 0.01-0.02%. According to recent information, it seems that in the USA the mortality rate of birds due to collisions at wind farms is negligible (Erickson et al., 2001). An exception is the golden eagle population at the Altamont pass. A comprehensive study of radio-tagged birds showed that in three years at least 20% of subadult birds and at least 15% of non-territorial adults were killed due to wind turbines. Juveniles (1% victims due to wind turbines) and breeding adults (4% victims due to wind turbines) were less strongly affected (Hunt, 2000). Other anthropogenic causes of death were considerably more important in the USA than wind farms (Tab. 11).

Table 11. Estimates of numbers of bird victims from collisions with anthropogenic structures in the USA (Erickson et al., 2001)

Cause	Estimated annual number of casualties
Traffic	60.000.000 – 80.000.000
Buildings and windows	98.000.000 – 980.000.000
Electricity pylons and cables	174.000.000
TV and communications towers	4.000.000 – 50.000.000
Wind turbines	10.000 – 40.000

In Germany, the proportion of wind farm victims ought to be higher because of the higher number of wind farms, but in Germany there are fewer vehicles, buildings, lead and radio and TV towers compared with the USA.

In Spain, wind turbines might have a particular impact on the mortality rate of griffon vulture. The death toll certainly runs into hundreds of victims per year (Lekuona, 2001; SEO, 1995). The Spanish population consists of approximately 8,100 breeding pairs, and represents the majority of the total European population of 9,300-11,000 breeding pairs (BirdLife International & European Bird Census Council, 2000).

In order to estimate the significance of the numbers of casualties in Table 10 for overall mortality rates, we looked at two examples. In Germany, there are approximately 12.000 breeding pairs of red kites and approximately 470 breeding pairs of white-tailed eagles. Taking into account that in addition there are juveniles and other non-breeding birds in the population, Germany probably supports approximately 36.000 individual red kites and approximately 1.400 individual white-tailed eagles. Assuming that 100 red kites are killed by wind turbines each year, then the additional mortality is 0.3% added to the yearly mortality rate. The number 100 might be unrealistic, because many red kites that are killed will not be found. Dead white-tailed eagles will certainly found more frequently. Assuming 10 victims in Germany each year, then the increase added to the background mortality rate will be approximately 0.7%.

The breeding populations in Germany of most of the other bird species listed as casualties in Table 10, are far higher than red kite and white-tailed eagle. Therefore- with the exception of gulls – it is not thought that wind turbines cause significant increases in annual mortality rates.

3.3.2 Collisions of bats with wind farms

Since the early 1960s it has been known, that bats could also be killed by wind turbines (Hall & Richards, 1962). However, only during recent years have studies been made of the scale of the mortality of bats due to wind turbines; like bird studies these have been carried out mainly in the USA. Tab.12 summarises the studies in which the collision rate per year (number of bats per turbine per year) was calculated. Even though there were clearly fewer data available than for birds, the results plot over a similar range. At some wind farms, only a few or even no losses were recorded, while at other wind farms large numbers of bats were killed.Another study about wind farms in Brandenburg determined an average collision rate of 0.23 bats per wind turbine per year, these values have not been corrected for search effort or rate of carcass scavenging (Dürr, 2003b).

For bats, there is an indistinct correlation between collision rate and height of the wind turbines (Fig. 17), but which is not statistically significant. Other studies also conclude that

Table 12. Collision rates of bats (annual number of victims per turbine) in different wind farms.

Country	Wind farm	Habitat	Collision rate	Comments	Source
Spain	Salajones	Mountain ridges	13.36		Lekuona, 2001
Spain	Izco-Albar	Mountain ridges	3.09		Lekuona, 2001
Spain	Alaiz-Echague	Mountain ridges	0		Lekuona, 2001
Spain	Guennda	Mountain ridges	0		Lekuona, 2001
Spain	El Perdón	Mountain ridges	0		Lekuona, 2001
USA	Buffalo Ridge	Grassland	2.3 Osborn et al. 1996		
USA	Foote Creek Rim	Prairie	1.34	Further studies in other years	Young et al., 2003a
USA	Vansycle	Farmland, Grassland	0.4		Strickland et al., 2001b
USA	Altamont	Mountain ridges	0.0035	Further studies in other years	Smallwood & Thelander, 2004
USA	Mautaineer Wind Energy Facility Blackwater Falls	Woodland	50		Boone, 2003
USA	Nine Canyon Wind Project	Prairie	3.21		Erickson et al., 2003
Australia	Tasmania	Coast	1.86		Hydro Tasmania

more bats are killed at wind farms with taller turbines (Dürr, 2003b). In Germany, according to Dürr (pers. comm.) no bats have been found at smaller wind turbines (<500 kW).

It was not possible to determine clearly whether wind farms are more dangerous in some environments than in others, because of the small number of studies. However, the high mortality rate at the only wind farm in a forest (Blackwater Falls, USA) was notable. A significant difference between the category „forest" and the remaining habitats does not exist (Kruskal-Wallis-Test, Chi²=2.57; df=1; not significant). There are also signs in Germany that bats are more endangered by wind turbines close to woods than by wind turbines in open habitats (Bach, 2002). Nathusius's pipistrelle, common pipistrelle and greater mouse-eared bat were found disproportionately frequently at wind farm locations close to trees and hedges. This does not apply to the noctule (Dürr, 2003b).

Numerous bat species were also identified among turbine collision victims. Further details of both bird and bat victims can be found in the comprehensive summary of Tobias Dürr at the Landesumweltamt Brandenburg (Dürr, 2003b; Dürr & Bach, 2004) (Tab.13).

All studies on bat collisions that were carried out over a sufficient period of time, showed that bats were killed predominantly in late summer and autumn, thus during their wandering and migrating phase (Dürr, 2003b; Keeley et al., 2001b). The species most affected are fast-flying and migrating ones (Dürr, 2003b; Johnson et al., 2003).

Table 13. Bat casualties at wind turbines in Germany. Total numbers found since 1998. Data from Staatliche Vogelschutzwarte, LUA Brandenburg, T. Dürr, 06.09.2004. BB = Brandenburg, ST = Sachsen-Anhalt, SN = Sachsen, TH = Thüringen, MVP = Mecklenburg-Vorpommern, SH = Schleswig-Holstein, NDS = Niedersachsen, NRW = Nordrhein-Westfalen, RP = Rheinland-Pfalz, HS = Hessen, SL = Saarland, BW = Baden-Württemberg, BY = Bayern.

Species		BB	ST	SN	TH	MVP	SH	NDS	NRW	RP	HS	SL	BW	BY	Total.
Noctule	Nyctalus noctula	40	1	20	54		3		1					1	120
Leisler's bat	Nyctalus leisleri	5	1	1	3										10
Serotine	Eptesicus serotinus	2			2		1		1						6
Particoloured bat	Vespertilio murinus	1		7											8
Greater mouse-eared bat	Myotis myotis			7											7
Daubenton's bat	Myotis daubentoni	1													1
Nathusius's pipistrelle	Pipistrellus nathusii	17	1	23	2				1						44
Common pipistrelle	Pipistrellus pipistrellus	15	2	6	2										25
Pipistrelle sp.	Pipistrellus sp.	4					14								18
Grey long-eared bat	Plecotus austriacus	1													1
Unidentified bats				2							2				4
Total		87	5	59	70	0	18	0	3	2	0	0	0	1	245

One possible reason for the collisions may be that migrating bats depend not only on ultrasound orientation, but also on other orientation techniques, and therefore do not notice the rotating blades (Johnson et al., 2003). Some locations where bat victims were found suggest that noctules are killed while trying to roost in the turbine nacelle. Pollution by gear oil from turbine machinery cannot be ruled out either (Dürr, 2003b), but according to more recent studies is unlikely (Dürr & Bach, 2004).

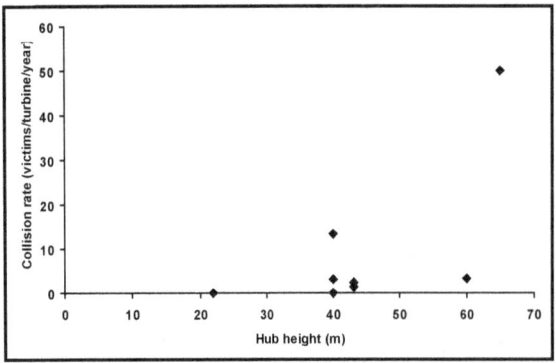

Figure 17. Collision rates of bats at wind farms of different tower heights.

4 Collision effects on population dynamics

The analyses so far show that collision rates of birds and bats at wind farms are generally small. Nevertheless, appreciable increases in mortality rates due to wind farms could occur at particular places. The impacts of increased mortality rates have so far only been measured directly in the field in one population, that of golden eagle in the Altamont area of California, USA (Hunt, 2002). During recent years, approximately 75-116 individuals have been killed (Smallwood & Thelander. 2004). Long-running studies of ringed and radio-tagged birds, as well as records of birds found dead at several wind turbines indicate that in spite of high losses the golden eagle population remained stable, but that numbers of subadults and non-breeding birds were reduced (Hunt, 2002). Continuing increases in mortality rates would lead to the probability that the population could not survive on its own; it would therefore be dependent on immigrants from outside - and a population decline can be predicted.

The following section tries to estimate the impact of additive direct losses due to wind farms on population development of different bird and bat species. Population models were used in order to investigate the extent to which an increase in mortality due to wind farms affects population growth. In order to estimate the degree of impact, two values have been chosen for turbine mortality, 0.1% and 0.5%. These values were added to the annual mortality rate. Even though they were chosen arbitrarily, they are within the range of the small number of available estimates (see chapter 3.3) and thus should give a realistic estimate of the overall impact.

4.1 Application of the population simulation model

The computer program VORTEX (version 9) was used to estimate population growth. This program is a tool for population viability analysis (PVA), is based on random simulations and was developed by the Chicago Zoological Society.

Various permutations of the population model can be saved as scenarios. Apart from population size and carrying capacity the type of reproductive system, average reproductive rate, mortality rate and the standard deviations of each are required to run the program. The removal of individuals for example through hunting (or wind energy), or supplementation of the population through releases or in-migration may also be included.

As a first step, data on population size, reproductive rate and mortality rate were taken from the literature for 23 birds and 4 species of bat (Tab. 14). With these data, scenarios were developed and the population trend calculated.

However, no population sizes were available in the literature for bats, and so a population size of 10.000 individuals was assumed. This approach had no major impacts on the subsequent calculations. The population size of some common bird species had to be reduced, so as not to exceed the program's maximum population size of 30.000 individuals. Both, standard deviations of values used for reproductive rates and mortality rates were required to simulate natural variations. Because standard deviations frequently could not be taken from the literature, their values were assumed to be 10% of the values given for mortality and reproductive rates. Density-dependant regulation of the reproductive rates was disregarded. Population development was examined over a period of 20 years and 100 iterations per scenario were run.

Following calculations using data precisely as given in the literature, these values were adjusted so that they resulted in a stable population development (Scenario 1) (Tab. 15). Then the mortality rate for all age groups was increased by an additional 0.1% (Scenario 2) and by an additional 0.5% (Scenario 3) and the resulting population development was calculated (Tab. 15). For white stork and black stork, age groups 2-3 and 3-4 were not considered, because birds of these ages do not occur in central Europe and thus hardly ever come in contact with wind farms.

Finally, for eight bird species, we estimated the extent to which reproductive rates must increase in order to offset additional mortality.

Table 14: Population parameters used for calculating population developments with VORTEX in Scenario 1

Species	Age of first breeding	Max age	Breeding system	Reproductive rate	SD	Proportion non-breeders	Mortality rate 0-1	SD	Mortality rate 1-2	SD	Mortality rate 2-3	SD	Mortality rate 3-4	SD	Mortality rate adult	SD	Immi-gration	References
Barnacle goose	2	21	Monogamous	0.48		0.1	34.9	6	12	3					12	3		Bezzel, 1985; Owen & Black,1989; Ganter et al.,1999; Ebbinge, 1991
White-fronted goose	2	17	Monogamous	0.66		0.1	40	5	15.5	3					15.5	3		Mooij et al., 1999
White stork	4	30	Monogamous	1.85		0.3	60	2.9	26.5	2.9	19.9	2.9	17.8	2.9	17.8	2.9		Burnhauser, 1983; NABU BAG Weißstorchschutz, 2004
Black stork	3	18	Monogamous	2.36		0.4	60	2.9	26.5	2.9	24.5	2.9			24.5	2.9		Bezzel, 1985; Möller & Norttorf, 1997
White-tailed eagle	4.4	36	Monogamous	1.35	Young/successful pair	0.25	30	5	17	3	17	3	17	3	17	3		Bezzel, 1985; Struwe-Juhl, 2002
Golden eagle	4	25	Monogamous	0.24		0.05	30	6	10	2	10	2	10	2	7.5	1.3	4 per year	Kostrzewa & Speer, 1995
Hen harrier	2.5	16	Monogamous	1.5	Young/successful pair	0.17	60	5	20	2					20	2		Koks et al., 2001
Kestrel	2	17	Monogamous	3.94	Young/successful pair	0.64	68	5	34	3					31	3		Village, 1990;Kostrzewa, 1993
Red kite	2	25	Monogamous	1.2		0.25	60	5	25	1.9					18	1.8		Bezzel, 1985; Kostrzewa & Speer, 1995
Crane	5	30	Monogamous	1.08		0.15	60	5	15	1.3	15	1.3	15	1.3	15	1.3		Prange, 1989
Corncrake	1	15	Monogamous	7.1		1.8	76.4	10							76.4	3		Green, 1999; Bezzel, 1985
Golden plover	1	12	Monogamous	1		0.1	53	4.4							22	1.8		Bezzel, 1985; Pearce-Higgins &Yalden, 2003
Lapwing	1.5	25	Monogamous	0.59		0.2	40.1	5.9							17.2	0.0	1	Catchpole et al., 1999; Peach et al., 1994
Curlew	3	31	Monogamous	0.57		0.1	40	4	18	1.8	12	1.2			12	1.2		Grant et al., 1999; Pearce-Higgins &Yalden, 2003
Black-tailed godwit	2	15	Monogamous	0.87		0.2	40	4	20	2					20	2		Struwe, 1995; Beintema & Müskens, 1981;Schekkermann & Müskens, 2000; Groen & Hemerik, 2002; Beintema & Drost, 1986
Oystercatcher	3	35	Monogamous	0.36		0.1	40	4	8.5	1	8.5				8.5	1		Bezzel, 1985; Durell, 2000; Goss-Custard et al. 1983
Redshank	1	16	Monogamous	1.43		0.3	55	5							31.5	4		Bezzel, 1985; Stiefel & Scheufler, 1984
Black-headed gull	1	26	Monogamous	1.25		0.3	56	5							27	3		Bezzel, 1985; Prévot-Julliard et al., 1998
Herring gull	5	33	Monogamous	1.15		0.3	78	6 Until sexual maturity							12	3		Bezzel, 1985; Wilkens & Exo, 1998
Skylark	1	10	Monogamous	1.45		0.6	50	5							35	3		Bezzel, 1993; Donald et al., 2002
Meadow pipit	1	8	Monogamous	4.26		2.93	74	7.2							54	5.2		Hötker, 1990
Starling	2	21	Monogamous	7.1		0.9	70	7.3	51	5					51	5		Bezzel, 1993; Glutz von Blotzheim, 1997
Corn bunting	1	8	Polygynous	1.93		1	47	4							39	4		Glutz von Blotzheim, 1997

Bats

Species	Age of first breeding	Maximum age	Breeding system	Reproductive rate	SD	Mortality 0-1 years	SD	Mortality Adult.	SD	References
Noctule	1 yo.	12 yo.	Polygamy	1.65	0.5	46	4	44	4	Krapp, 2004
Serotine	1 yo.	12 yo.	Polygamy	0.5	0.2	39	3.9	11.5	1.2	Krapp, 2001
Common pipistrelle	1 yo.	16 yo.	Polygamy	0.95	0.2	50	5	23.5	2.5	Krapp, 2004
Nathusius's pipistrelle	1 yo.	11 yo.	Polygamy	1.1	0.3	45	4.5	45	4.5	Krapp, 2004

Table 15. Results of model calculations of population developments of selected bird and bat species under different scenarios

Species	Estimated starting population	Population at end of Scenario 1	Population at end of Scenario 2	% of population at end of Scenario 1	Population at end of Scenario 3	% of population at end of Scenario 1
Barnacle goose	11000	12821	12105	94.42	10939	85.32
White-fronted goose	4000	3918	3778	96.43	3296	84.12
White stork	23310	21283	21262	99.9	19424	91.27
Black stork	1746	1957	1927	98.47	1782	91.06
White-tailed eagle	1482	1825	1762	96.55	1616	88.55
Golden eagle	124	132	129	97.73	122	92.42
Hen harrier	618	668	649	97.16	569	85.18
Kestrel	14000	15376	14867	96.69	13096	85.17
Red kite	28260	27151	25964	95.63	23311	85.86
Crane	6300	6658	6504	97.69	5772	86.69
Corncrake	2680	3317	3073	92.64	2774	83.63
Golden plover	20000	20431	19160	93.78	17960	87.91
Lapwing	10000	10355	10172	98.23	9173	88.59
Curlew	8800	8670	8449	97.45	7810	90.08
Black-tailed godwit	14000	13269	12998	97.96	11555	87.08
Oystercatcher	6000	5951	5841	98.15	5370	90.24
Redshank	2600	2749	2629	95.63	2168	78.87
Black-headed gull	5800	6126	5810	94.84	5126	83.68
Herring gull	9000	9876	9186	93.01	8101	82.03
Skylark	7200	8293	7506	90.51	6377	76.9
Meadow pipit	3000	2882	2775	96.29	2229	77.34
Starling	6800	7839	6764	86.29	6108	77.92
Corn bunting	3300	3456	3450	99.83	2811	81.34
Bats						
Noctule (Nyctalus noctula)	10000	10701	10379	96.99	9393	87.78
Serotine (Eptesicus serotinus)	10000	10782	10742	99.63	9565	88.71
Common pipistrelle (Pipistrellus pipistrellus)	10000	9677	9365	96.78	8593	88.80
Nathusius's pipistrelle (Pipistrellus nathusii)	10000	10610	9980	94.06	9090	85.67

Scenario 1	Original values
Scenario 2	Increase of mortality by 0.1%
Scenario 3	Increase of mortality by 0.5%

4.2 Results of the population simulations
Birds

The impact on population development of additional mortality due to wind farms varies greatly between species. However, even with relatively small increases in mortality, decreasing trends are clearly recognisable for all species. The largest population declines are shown by species which start breeding at one year old (Fig. 17). The influence of additional mortality decreases for birds which start breeding for the first time up to the age of four, but increases again for those that first breed aged five. Likewise, additional losses have a stronger negative effect on the population size of short-lived species than on long-lived species (Fig. 18 and 19). These differences are particularly evident in Scenario 3.

If the percentage population decrease is compared with adult mortality, it is noticeable that species with high mortality rates show a stronger negative reaction to additional losses than species with low mortality rates. In summary, relatively long-lived species with low adult mortality show smaller population decline than short-lived species.

Following this, estimates were made of the extent to which losses caused by wind farms can be compensated for by an increased reproductive rate (Tab. 16).

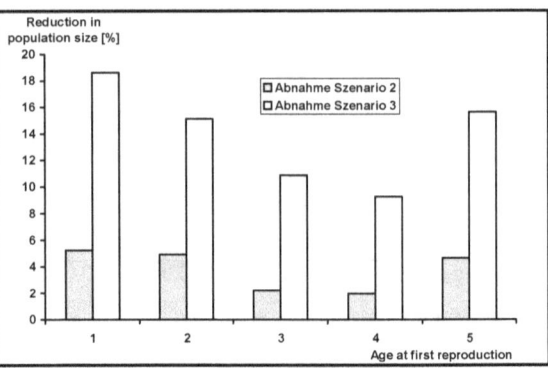

Figure 18. Results of population modelling with VORTEX. Reductions of population size in relation to age of first reproduction.

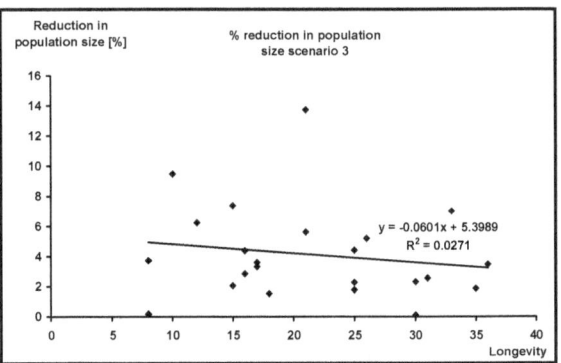

Figure 19. Population developments of different bird species (dots) under scenario 2 (increase of annual mortality rate by 0.1%) in relation to longevity.

Table 16. Substitution of increased mortality rates of different bird species by increased reproduction rates under different scenarios

Species	Age of first breeding	Maximum age	Increase (%) in reproductive rate under Scenario 2	Increase (%) in reproductive rate under Scenario 3
Skylark	1	10	0.8	2.1
Meadow pipit	1	8	0.001	1.5
Golden plover	1	12	0.3	2.3
Lapwing	1.5	25	0.7	3.4
Oystercatcher	3	35	0.8	6.9
White stork	4	30	0	5.4
White-tailed eagle	4.4	36	1.6	5.9
Crane	5	30	1.9	7.4

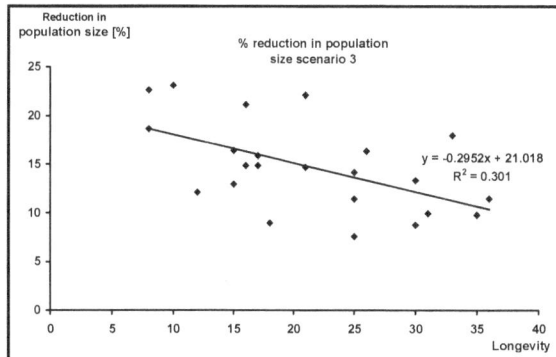

Figure 20. Population developments of different bird species (dots) under scenario 3 (increase of annual mortality rate by 0.5%) in relation to longevity.

Short-lived species that are already sexually mature at one year old, require considerably smaller increases in reproductive rates in order to offset the additional mortality than long-lived species reaching sexual maturity at a later age.

Bats

Bat species differ only slightly in their population-biological data (Tab.14). However, taking into account adult mortality and reproductive rate, two groups can be formed: one group consists of noctule and Nathusius's pipistrelle, the other of serotine and common pipistrelle. Both groups show different rates of population decline due to increased mortality caused by wind farms. In comparison to birds, the population decline of bats is mostly lower.

4.3 Discussion of the simulations calculations

Sensitivity analyses, like the ones presented here, could identify impacts of additional losses on population development. The models are very simple and consider only some factors with an influence. Wind farms also could have additional impacts on bird populations by causing displacement.

Other studies (Dierschke, Hüppop & Garthe, 2003) assume increases in mortality rates of 5% (recommended by NERI 2000). This calculation determined that the 5% increase in mortality due to wind farms has a relatively higher impact on species with high mortality rates than on species with lower mortality rates. For example, in our calculation for 100 white-tailed eagles, instead of 17 birds now 17.85 individuals would be killed each year. Instead of 54 meadow pipits, now 56.7 individuals would die. The impact on the meadow pipit population is therefore higher than for white-tailed eagle. We did not increase the mortality in this way, because we assumed that a constant proportion of the population (0.5% and 0.1%) was killed through wind turbines each year.

In our models, short-lived species generally reacted with larger percentage population decreases than long-lived species. However, short-lived species can better compensate for these losses, because a smaller percentage increase in reproductive rate is needed than for long-lived species. In principle, our results show similar trends to those of Dierschke et al. (2003).

Morrison et al. (1998) compared the sensitivity of different bird species groups with the mortality rates of different age groups. Songbirds reacted more strongly to changes in the mortality of young birds than to that of adults, while ducks reacted equally strongly to changes of mortality of both youth birds and adults. Geese, gulls and eagles are more sensitive to changes in adult mortality than that of young birds.

Morrison & Pollock (1997) determined that increased mortality of young birds can more easily be compensated for by increased reproductive rates, than increased adult mortality. Therefore, our results that losses of short-lived species can be readily compensated for by increased reproductive rates, are easily explained. Increased reproductive rates should actually be expected in certain circumstances. For species with populations at carrying capacity, reproductive rates are limited by competition and density; but these would rise if mortality rates increased. Nevertheless, for species whose population size is not at carrying capacity or whose reproductive rates are limited due to other factors, e.g. habitat quality or climate factors, it is impossible to compensate for additional losses due to wind farms.

Additional losses have a smaller effect on the population dynamics of bats than of birds, as bats have a polygamous reproductive system. Therefore the loss of a female should have a stronger effect than that of a male, but no account could be taken of this in the modelling. Nevertheless, it is reasonable to assume an equal distribution of female and male losses.

The model calculations presented here are very simple, because they assume an equal likelihood for all species of individuals being killed at wind farms. In reality this is unlikely to be the case. For example, more white-tailed eagles are reported killed at wind farms than white storks, even if in theory equal numbers should be reported for each species. The white stork population is barely larger than the white-tailed eagle population. Possible reasons for the different levels of risk can only be guessed at. Still, the example makes clear that for each bird species a specific risk assessment should be carried out in order to adjust the model calculations to reality. These specific assessments also have to consider other cumulative risks. A population viability assessment should also be used to determine whether cumulative mortality can be offset by increased reproductive rates.

It should be borne in mind that mortality due to wind energy is probably less selective than natural causes of death. Not only the least fit individuals (in a population genetics sense) are killed, but also individuals which are potentially important for population development, as shown by the example of red kite. On the other hand, it cannot be ruled out that losses due to wind farms are compensated for by reduced mortality due to other causes.

It is fundamentally not possible to say whether the loss of an individual of a short-lived species with a high reproductive rate (such as many species of songbird) is less serious than the loss of an individual of a long-lived species. It is harder for a long-lived species to offset mortality through increased rates of reproduction. In general, long-lived species (often large birds or sea birds) have distinctly smaller populations than short-lived songbirds. A collision of an individual of a long-lived species results in a larger increase in the overall mortality rate and therefore has a relatively stronger impact on the population than the collision of an individual of a short-lived species, such as a species of songbird.

5 Measures to reduce the impacts of wind farms

Research into methods of reducing mortality of birds and later bats has been initiated, in particular in the USA (Smallwood & Thelander, 2004; Sterner, 2002; US Fish and Wildlife Service, 2003). Many recommendations have arisen from these studies, which are summarised here, supplemented with results from other available literature. The negative influences, such as displacement and barrier effects, are also taken into consideration. The recommendations can be roughly divided into: site selection; managing habitat within the wind farm; configuration of wind turbines within the wind farm; operation of the wind farm; and other site-specific measures to be carried out at individual wind farms.

5.1 Choice of site

Despite all studies on measures to be taken in and around wind farms, the choice of the right site is still the most important method to reduce the negative effects wind farms have on birds and bats. Compiled annual collision rates of birds and bats (Tables 9 and 12) show that sites in areas with a high occurrence of birds of prey (mountain ridges), as well as wetlands and forests (bats) should not be chosen as wind farm sites. In the USA, recommendations exist for the choice of a wind farm site (US Fish and Wildlife Service, 2003):

- avoidance of sites with protected animals and plants;
- avoidance of sites with sensitive bird species (5km distance from the display sites of prairie hen);
- avoidance of sites, which are well-known as migration routes, flight corridors, or where birds are highly concentrated for other reasons;
- avoidance of well-known sites for hibernating, breeding or migrating bats;
- avoidance of locations with high occurrence of birds of prey (mountain ridges, area with high densities of prey);
- avoidance of habitat fragmentation by wind farms (wind farm should not fragment coherent habitats)

Other authors also recommend avoiding critical sites (mountain ridges, wetlands, forests (Sterner, 2002; Strickland et al., 2001a)). According to current knowledge, the hunting areas of the serotine bat, which are within a 200m radius of woods or particularly insect-rich sites, should also be kept free of wind farms to avoid losses of hunting ground (Bach, 2002; Rahmel et al., 2004). In order to protect birds of prey, it is furthermore recommended in certain areas avoid mountain ridges by at least 50m (Hoover, 2002; Johnson et al., 2000). Because in some cases only a few wind turbines are responsible for the majority of losses (for example at Altamont), it would be worth considering to removing these turbines (Sterner, 2002).

Important roosting areas for waders and water birds should also be kept free of wind farms. A buffer distance of at least 400m is recommended, and for goose roosts at least 500m. These values apply to most wind turbines currently in operation (hub height under 50m), but need to be confirmed for future, taller wind turbines. Well-known migration and flight corridors should be kept free of wind farms.

5.2 Design of the environment around wind farms

Studies on different wind farms in the USA, but also in Germany, indicate that the high number of bird of prey victims could be explained by their being attracted to the environment surrounding the wind farms. This can happen either because there is a high density of food already, or because the food density is increased during wind farm construction. Such development of the area could produce habitats, for example fallow land, which are advantageous for small mammals, the main food source of many birds of prey. Perching places, such as fences but also lattice-towers on some sites, which act as additional attractions to birds of prey, could also result from wind farm development.

The recommendations are accordingly:

- avoidance of features, which could lure birds and bats (ponds, habitat edges, areas with a high density of small mammals, e.g. fallow land, etc.) (Bach, 2003; Hoover, 2002; Kelly, 2000; Rahmel et al., 2004; Sterner, 2002; US Fish and Wildlife Service, 2003);
- minimising infrastructure as roads, fences etc, avoid creating and consider removing perches (Kelly, 2000; Sterner, 2002; Strickland et al., 2001a; US Fish and Wildlife Service, 2003);
- Removal of carcasses (to minimise attraction of birds of prey) (US Fish and Wildlife Service, 2003)

Deliberately chasing birds away from the wind farm area has also been considered (Kelly, 2000); see more below.

5.3 Configuration of wind turbines within a wind farm

A number of studies demonstrate that wind farms, which are arranged perpendicular to the main flight direction, have a stronger barrier effect and possibly cause more frequent collisions than wind farms running parallel to the lines of flight (Everaert et al., 2002; Isselbächer & Isselbächer, 2001). Accordingly, the recommendations are to arrange turbine arrays parallel to and not across the main flight direction. In addition, arranging the turbines into blocks, so that corridors are left, which could be used by birds as a safe passage is recommended (Albout et al., 1997; Albouy et al., 2001; US Fish and Wildlife Service, 2003).

5.4 The operation of wind farms

If collision problems are known to occur only at particular times of the year, which is often the case for bats, it is recommended that wind turbines are switched off during the main flight times (Kelly, 2000; Sterner, 2002; US Fish and Wildlife Service, 2003).

5.5 Design of individual wind farms

It is known from studies of various design characteristics of wind farms, that these can influence collision risk.

Tower construction

Only closed towers with no perching opportunities for birds of prey should be used (US Fish and Wildlife Service, 2003). Most studies show that lattice towers are dangerous (Sterner, 2002), see also Thelander & Rugge (2000). Particularly hazardous are the guy ropes used on older wind turbines and should not be used. Taller towers could be advantageous; wind turbines with particularly high mortality should be replaced by modern ones (repowering, see also chapter 6). If possible, the height of the mast should be selected so that collision rates are minimised (US Fish and Wildlife Service, 2003).

Connection to the transmission grid

Grid connection should be made under ground to avoid collisions with overhead electric cables (Albouy et al., 1997; Albouy et al., 2001; US Fish and Wildlife Service, 2003).

Lighting of wind farms

The risk of bird collisions with offshore oil production platforms is considerably increased by lighting (Marquenie & van de Laar, 2004). The highest number of bird victims in one night so far occurred at a lit, single turbine (Karlsson, 1983). Thus, light obviously attracts birds and increases the danger of collisions at night. Early studies assumed that the orientation of birds is more strongly influenced by white and red light than by green and blue (Poot, 2004). Because of the risk that birds are attracted to the flashing red lights used for safety lighting, the intensity of light should be reduced to a minimum and, if possible, the intervals between each flash should be made as large as possible. Instead of red light, a Strobo-Light is recommended, which attracts less birds (Sterner, 2002; US Fish and Wildlife Service, 2003). However, no studies into wind farm lighting to date have arrived at safe conclusions.

Increased visibility of turbine blades

Birds, which are very close to wind turbines, cannot see rotating turbine blades as solid objects, but only register them as a „motion smear". The distance at which this phenomenon occurs is roughly 20m for small, fast-rotating rotors and 50m for larger ones. This could be one reason for many collisions of birds of prey, which almost all happen during the day, and so at a time when their vision should actually function well. Neuro-physiological experiments determined that painting one of the three turbine blades black, or marking it with a black pattern, would increase its delectability. The mark has to be perpendicular to the rotor axis. Marking the ends of the rotors increases their delectability from the sides (Hodos, 2001; Hodos et al., 2001; McIsaac, 2001).

So far, field studies have not been able to determine the effectiveness of marking turbine blades. However, it should be noted that this is quite difficult to achieve, due to methodological reasons (Erickson et al., 1999; Sterner, 2002).

No measurable results have been reached so far from experiments coating wind turbines with UV-reflecting paint in order to reduce collisions, or to keep birds away from the danger zone (Strickland et al, 2001a; Young et al., 2003b).

The influence of rotor speed on collision risk has not yet been studied (Sterner, 2002).

It might be possible to make wind turbines more noticeable by acoustic signals, for example using a high-pitched whine (Dooling & Lohr, 2001), or warning signals (Sterner, 2002). However, such measures could be very dangerous for bats, which might get attracted to such signals (Bach in litt.).

Investigation is needed to determine if bats can be induced to use echolocation close to wind turbines.

5.6 Measures transferred on the conditions/circumstances in Germany

The recommendations listed so far in this chapter with the aim of decreasing negative impacts of wind farms on birds and bats are generally also applicable to Germany. Some problems (bird of prey mortality on mountain ridges and plateau edges) have not yet been recorded in Germany. For some specific problems, which have occurred in recent years in Germany, no solutions advice can be found in the literature. This is particularly true for the increasing number of collisions involving birds of prey, especially red kites and white-tailed eagles, in recent years. These collisions are not concentrated in particular places, but appear to be distributed randomly.

The degree to which measures to increase the detectability of turbines are able to solve problems is questionable. Measures to deter birds completely from a wind farm area should only be used if high rates of collision are expected. Such measures could be perhaps counter-productive for those wind farms, of which the negative effects include displacement of birds from their roosting places.

6 Estimation of the impacts of 'repowering'

Because, firstly, wind farm sites are becoming less available, particularly in North Germany, and also because wind turbine technology has made rapid progress in recent years, there is a trend to replace numerous small turbines with smaller numbers of larger ones (repowering). In all likelihood, this trend will continue over the next few years. Using the results collated so far, this chapter tries to evaluate the potential impacts of repowering on birds and bats. Impacts of disturbance-displacement and collision risk are considered. Four scenarios were modelled as a first step towards understanding the impacts of repowering:

Scenario 1 assumes that ten 0.15MW turbines are replaced by a single 1.5MW turbine. Taking into account the relationship between hub height and capacity (chapter 3.1.2), this implies that on average ten masts, each 27.4m high, are replaced by one 78.5m high. Scenario 2 assumes that three 0.5MW turbines are replaced by a single 1.5MW turbine; therefore on average three 47.4m high masts are getting replaced by one 78.5m high (chapter 3.1.2). These two scenarios assume no increase in capacity, which in practice happens only rarely. Therefore, in the other scenarios we raised the overall capacity of a wind farm by a factor of 1.5 (Scenario 3) and by a factor of 2 (Scenario 4). At the same time 0.5MW turbines were replaced by 1.5MW turbines.

The selection of these scenarios is based on the following consideration: impact data for wind turbines with a capacity greater than 1.5MW are rare. If larger turbines were chosen for the scenarios, then the results would be an unsafe extrapolation. Replacement of 0.5MW turbines with 1.5MW turbines combined with an increase in wind farm capacity of 1.5 or 2-times the total output is realistic (P. Ahmels, Bundesband Windenergie, pers. comm.). The replacement of both 0.15MW turbines and 0.5MW turbines with 1.5MW turbines with no increase in capacity (Scenarios 1 and 2) should allow an evaluation of the influence of wind turbine size.

6.1 Repowering and disturbance of birds

The relationship between hub height of wind turbines and the minimum distances birds are found from wind turbines, combined with the relationship between hub height and capacity, allow estimates to be made of the effect of repowering on the spatial distribution of birds. Hence, the following simple hypothesis is assumed:

1. No birds use the area within the radius of the minimal distance (disturbance area); outside this circle turbines have no effect.
2. It is a single, standalone wind turbine. Because wind farms differ greatly in layout, its effect cannot be generalised. The effect of layout must be considered separately for each wind farm.

An estimate of the impacts can only be made by comparing the size of the disturbed areas. If, for example in Scenario 1, the disturbed area of a single 1.5MW turbine is less than that of ten 0.15MW turbines, then repowering is recommended in order to reduce displacement disturbance, but otherwise it is not. The same applies to Scenario 2, so that repowering has a positive effect if the disturbed area of a single 1.5MW turbine is smaller than that of three 0.5MW turbines. For Scenario 4 the question is whether the disturbed area of two 1.5MW turbines is less than that of three 0.5MW turbines.

The results for each scenario are shown in Table 17. The results of Scenarios 1 and 2 do not differ from each other. When capacity remains the same, repowering has a positive effect on breeding birds. The picture is not clear for non-breeding birds.

Negative impacts were predicted for buzzard, kestrel, lapwing, black-headed gull and carrion crow. Under Scenarios 1 and 2, repowering would reduce the disturbed area for all other species. Buzzard, kestrel, black-headed gull and carrion crow are all species, which are not greatly displaced by wind turbines, so that even after being increased, the disturbed area is still quite small. Larger wind turbines should also have a negligible effect and so the impacts of repowering could be assessed as non-significant. Lapwing, however, show a strong reac-

Table 17. Assessment of the extent of the area in which disturbance of birds occurs after repowering old wind farms. Results of model calculations under different scenarios. „Positive" means a smaller area of disturbance after repowering, „negative" means a larger area after repowering. See text for details.

Scenario		1	2	3	4
Increase in wind farm capacity		No	No	1.5 x	2.0 x
Change in turbine size		0.15 MW to 1.5 MW	0.5 MW to 1.5 MW	0.5 MW to 1.5 MW	0.5 MW to 1.5 MW
Species					
Breeding season					
Mallard	*Anas platyrhynchos*	Positive	Positive	Positive	Positive
Oystercatcher	*Haematopus ostralegus*	Positive	Positive	Positive	Positive
Lapwing	*Vanellus vanellus*	Positive	Positive	Negative	Negative
Black-tailed godwit	*Limosa limosa*	Positive	Positive	Positive	Negative
Redshank	*Tringa totanus*	Positive	Positive	Positive	Positive
Skylark	*Alauda arvensis*	Positive	Positive	Positive	Positive
Meadow pipit	*Anthus pratensis*	Positive	Positive	Positive	Positive
Yellow wagtail	*Motacilla flava*	Positive	Positive	Positive	Positive
Blackbird	*Turdus merula*	Positive	Positive	Positive	Positive
Willow warbler	*Phylloscopus trochilus*	Positive	Positive	Positive	Positive
Chiffchaff	*Phylloscopus collybita*	Positive	Positive	Positive	Positive
Sedge warbler	*Acroc. schoenobaenus*	Positive	Positive	Positive	Positive
Reed warbler	*Acrocephalus scirpaceus*	Positive	Positive	Positive	Positive
Marsh warbler	*Acrocephalus palustris*	Positive	Positive	Positive	Positive
Whitethroat	*Sylvia communis*	Positive	Positive	Positive	Positive
Reed bunting	*Emberiza schoeniclus*	Positive	Positive	Positive	Positive
Linnet	*Carduelis cannabina*	Positive	Positive	Positive	Positive
Non-breeding					
Grey heron	*Ardea cinerea*	Positive	Positive	Positive	Positive
Wigeon	*Anas penelope*	Positive	Positive	Positive	Positive
Geese		Positive	Positive	Negative	Negative
Mallard	*Anas platyrhynchos*	Positive	Positive	Positive	Positive
Diving ducks		Positive	Positive	Positive	Positive
Buzzard	*Buteo buteo*	Negative	Negative	Negative	Negative
Kestrel	*Falco tinnunculus*	Negative	Negative	Negative	Negative
Curlew	*Numenius arquata*	Positive	Positive	Positive	Negative
Oystercatcher	*Haematopus ostralegus*	Positive	Positive	Positive	Positive
Lapwing	*Vanellus vanellus*	Negative	Negative	Negative	Negative
Common snipe	*Gallinago gallinago*	Positive	Positive	Positive	Positive
Golden plover	*Pluvialis apricaria*	Positive	Positive	Negative	Negative
Woodpigeon	*Columba palumbus*	Positive	Positive	Negative	Negative
Black-headed gull	*Larus ridibundus*	Negative	Negative	Negative	Negative
Starling	*Sturnus vulgaris*	Negative	Negative	Negative	Negative
Carrion crow	*Corvus corone*	Negative	Negative	Negative	Negative

tion regarding the size of a wind turbine. It can be assumed therefore that non-breeding lapwings are negatively influenced by repowering.

Scenarios 3 and 4, which are more realistic, present a different picture. Regarding breeding birds, for black-tailed godwit and above all for lapwing, negative impacts are predicted. For non-breeding birds, in addition to the species affected in Scenarios 1 and 2, other species which are sensitive to turbines (geese, waders of open habitats) also show negative effects in Scenarios 3 and 4.

6.2 Repowering and collisions of birds and bats

The relationship between hub height and collision rate of birds and bats (chapter 6.1) estimates to be made of the extent to which collision rates would change in each of the scenarios. If, for example, in Scenario 2 a 0.5MW turbine was replaced by a 1.5MW turbine, then, following the regression equations for the data in Fig.16 from chapter 3.2, the collision rate for birds would increase from 13.8 to 23.1. This is equivalent to a factor of 1.7, and therefore less then the factor of 3, which is the amount by the number of turbines has been reduced. Hence, with regard to birds, repowering under Scenario 2 can be assessed as having a positive effect. The corresponding values for bats are 10.9 victims per year for a 0.5MW-turbine and 37.6 victims per year for a 1.5MW-turbine. The values differ by a factor of 3.5, thus more than the factor of 3 by which the number of turbines is reduced. Therefore, it is predicted that repowering under Scenario 2 results in more bats being killed than previously.

The results of the calculations are presented in Table 18. By comparison with birds, bats are more sensitive to repowering than birds. For birds, increasing turbine size while keeping wind farm capacity constant seems to have little effect on collision risk. Increased capacity would, however, increase levels of risk. The critical point of the capacity increase clearly lies between a factor of 1.5 and 2.

It should be remembered that these modelled scenarios are based on only a few analysed cases and that the relationship between hub height and collision rate for birds and bats is not statistically proven. Moreover, the studies showing very high collision rates of bats and birds were often carried out in other countries and not in habitats or conditions comparable with those in central Europe.

Therefore, the presented results are at best preliminary estimates, and should be used with great caution. Even one new study could significantly change the main results, which therefore can in no sense be considered robust. In addition, it is not possible to estimate the risks posed by significantly greater size of the new generation of wind turbines on bats and migrating birds – particularly nocturnal migrants.

6.3 Summary of the assessment of repowering

If one understands by repowering the replacement of small, older wind turbines by a smaller number of larger turbines with no change in overall capacity, then on the basis of data currently available, positive impacts predomi-

Table 18. Assessment of collision rates of birds and bats in relation to repowering old wind farms. Results of model calculations under different scenarios. „Positive" means a smaller annual collision rate per turbine after repowering, „Negative" means a higher annual collision rate per turbine after repowering. See text for details.

Scenario	1	2	3	4
Increase in wind farm capacity	No	No	1.5 x	2.0 x
Change in turbine size	0.15 MW to 1.5 MW	0.5 MW to 1.5 MW	0.5 MW to 1.5 MW	0.5 MW to 1.5 MW
Birds	Positive	Positive	Positive	Negative
Bats	Positive	Negative	Negative	Negative

nate. A smaller number of larger turbines probably displace significantly fewer birds than a large number of smaller turbines. Even thought the collision rate of birds clearly increases to a certain extent with increased turbine size, this increase should probably be more than offset by the reduction in turbine numbers. This applies to birds and to a lesser extent also to bats.

If in the process of repowering the overall capacity of a wind farm is increased, then the advantages of repowering decrease. Particularly sensitive bird species are increasingly disturbed and the collision rate for bats increases. After carefully taking into consideration all of the estimates and partial results, the point at which negative impacts start to predominate is at around a capacity increase of 1.5 times (assuming that 0.5MW turbines are replaced by 1.5MW turbines).

The impact of very large and compulsorily illuminated wind farms on nocturnal migrant birds is largely unknown. So far, no concrete evidence exists for bird victims. However, note that in Germany systematic searches for corpses have been carried out at only a few wind farms, compared with the USA, where such monitoring is standard (Morrison, 1998; Morrison & Pollock, 2000). In particular, small, inconspicuous passerines, which make up a large proportion of nocturnal bird migrants, are likely to be overlooked if they collide with wind turbines.

However, repowering also offers important opportunities. Wind farm sites which have adverse effects on birds and bats could be given up, and replaced with new ones constructed on less problematic sites. Such a process of „land consolidation" for wind power could resolve many conflicts between nature conservation and wind energy development.

7 Impacts of other types of renewable energy

There are a number of renewable energy technologies in addition to wind power. These are in particular hydroelectric power, solar energy and the cultivation of „energy crops". Hydroelectric power has been used as an energy source for a long time. The effects on the biological environment are complex (Umweltbundesamt, 2001). The effects of hydropower on birds and bats are mainly indirect. On the one hand, the installation of dams and impoundments create new bodies of still water, which offer new habitats for a number of species, while on the other hand there will be impacts on flowing water and meadow habitats, as well as on the natural dynamics of river systems. Because the effects of hydropower on other taxa (for example migratory fishes and riverine insects) are incomparably larger than on birds and bats (Bunge et al., 2001; IKSR et al., 2003; Meyerhoff et al., 1998), hydropower is not considered further in this report.

Solar energy

The usage of solar energy for generating electricity has recently undergone a rapid increase. Although solar cells (both for electricity generation and heating water) have so far mainly been installed on buildings, now solar parks covering several hectares are starting to come into operation. The impacts on birds and bats of such installations, where solar panels are installed on frames on uncultivated land, are almost completely unknown. There are two possible impacts, analogous to those at wind farms: (1) displacement of breeding and non-breeding birds, as well as bats, from the area of the solar parks; and (2) collision mortality.

Birds and less so bats could become collision victims, because the more or less reflective surfaces of the solar cells could imitate areas of water, to which the birds are attracted. This phenomenon occurs regularly on roads after falls of rain, and water birds are caught in this trap. Potential victims are waders and waterfowl, which mainly migrate at night. At present there is no indication as to whether such collisions are significant and therefore further research is needed (see chapter 8).

Impacts of large installations of solar panels on breeding birds cannot yet be predicted. However, one can assume that sensitive species of open habitats (breeding waders) will not remain in solar parks. It is not possible to draw any general conclusions on the extent to which breeding songbirds are affected. Crucial to this will be the management of land between arrays of solar panels. The same applies to non-breeding birds, so that geese and sensitive waders are unlikely to be found within the solar parks. In order to assess the impacts from the nature conservation point of view, it is essential to know what was the management of the land between the solar panels before they were installed.

Energy crops

Currently there are two main potential ways of exploiting plants as energy crops: the extraction of oil for fuel and using whole plants, either for direct combustion, or for fermentation.

Oil crops

Fuels are already extracted from crops, principally oil-seed rape and sunflowers. The cultivation of these plants for energy use is no different from their cultivation for other purposes, so that the experience from agriculture is highly relevant. The intensive cultivation of oil-seed rape is in general no different in its effects on bird populations than other intensive forms of agriculture. As a winter-sown crop, with a more-or-less complete surface cover, even in winter, compared with winter cereals it provides food for some birds, particularly large, wintering species (swans, geese, wigeon, great bustard) outside the breeding season (Gillings, 2001; Inglis et al., 1997). Because the rapidly growing rape plants grow very close together in spring, farmland birds such as grey partridge, quail, lapwing, meadow pipit, corn bunting and yellowhammer avoid them (Biber, 1993; Döring & Herfrich, 1986; Fuchs, 1997; Jenny et al., 2002; Morris et al., 2001; Salek, 1993; Sellin, 1994; Suter, Rehsteiner & Zbinden, 2002; Töpfer, 1996; Wakeham-Daeson & Aebischer, 1997; Weibel, 1995). In particular second or replacement broods are barely possible, because birds, which have settled in the short plants in spring, become caught in an „ecological trap" later in the season (Donald et al., 2001; Wilson et al., 1997). In other regions, though, some bird species are relatively frequent, such as reed bunting (Burton et al, 1999), especially yellow wagtail and bluethroat in North Germany (own observations) and Poland (Stiebel, 1997; Tryjanowski & Bajczyk, 1999), but not obviously so in Great Britain (Mason & Macdonald, 2000). Less evidence is available for interactions between cultivation of sunflowers and bird populations. Skylarks avoid sunflower fields during the breeding season (Weibel, 1999; Weibel, 1995), but depending on the wild weeds which are present, marsh warbler and whitethroat might move into the field from the edge (Dürr in litt.). Fields of sunflower-stubble are valuable food sources for granivorous songbirds and for pigeons.

It should be borne in mind that several farmland bird species require a minimum degree of diversity in countryside management (NABU, 2004). If the cultivation of oil-seed rape and sunflowers for oil production causes a widespread standardisation of crop structure, negative impacts on bird populations are expected.

Woods

Fast-growing trees, in particular willows and poplars, are cultivated in several countries for wood fuel. The tree crop will be harvested after three to five years, when the stems are approximately 5m high. About six crops (so 20-30 operational years) are possible per cultivation area. Only a few studies have been carried out so far (Anderson, Haskins & Nelson, 2003; Goransen, 1990; Kavanagh, 1990; Sage & Robinson, 1996) and these show that areas planted with trees are settled by typical farmland birds in their first year of growth and then increasingly by birds which typically breed in scrub. The density of bird populations depends on a variety of factors, among others how strictly the necessary weed control has been carried out during the first years of growth.

Other energy crops

Little information is available on the impacts of other plant-based energy sources on bird and bat populations (Anderson et al., 2003). Very tall plants such as *Miscanthus* should be unsuitable as breeding and roosting habitats for farmland birds, but should attract inhabitants of reeds as breeding and roosting species. Studies in North Germany illustrated that certain ways of growing grass show a higher density and variety of bird species than conventional arable fields (Beyea, Cook & Hoffman, 1995). It is possible to obtain bio-gases from agricultural crops. Plants high in protein, fat or carbohydrate are most suitable. Crops such as silage (first cut), maize and oil-seed rape (www.fnr-server.de) are examples, while products which develop from a more extensive, nature-orientated regime (late cut grass), are less suitable. Farmland with fast-growing maize species is probably uninhabitable for most farmland birds, but could be used by ground-breeding species shortly after sowing. A distinct increase in maize cultivation could cause a considerable additional decline in farmland bird diversity. This is also expected, if „energy plants" take up large areas as monocultures and consequently the habitat edges, which are important for many life forms, are lost. At present, many bird species are able to survive in arable land only at the edges and not in the actual production areas (NABU, 2004).

8 Research requirements

Despite the evidence gathered over the last two decades on the impacts of wind farms on birds and to a lesser extent on bats, there are still serious gaps in our knowledge, which, with regard to further extension of wind energy, and in particular of repowering, need to be resolved urgently. Here, the special emphasis lies in the aspects, which are relevant to the further extension of wind energy in Germany and from which practical experience are expected relevant to future planning. With regard to other forms of renewable energy production, research is still almost in its infancy, and this area has to be dealt with separately.

Research requirements on the impacts of wind farms in birds and bats

Many studies in Germany and other European countries have examined the displacement effect caused by wind farms. As shown in chapter 2 „materials and methods" only a few studies so far fulfil the criteria required for well-founded research. In the case of wind power this implies that many more before-after studies are necessary covering at least two years each before and after the installation of the wind farm and also that these should include a control area with no wind farm development. For statistical reasons, a minimum of two research years before and after construction are necessary in order to take account of the natural variations in populations (or in any other analysed parameter).

The influence of wind energy development on habitat choice of birds and bats has not been explicitly considered by many studies. Detailed analyses are required, in particular for species sensitive to disturbance (roosting geese and waders). These studies should consider equally the extent of the available habitats, their usage by birds and the resultant interactions with the wind farms. The results of such studies should allow for much more elaborate assessment of separation distances of bird roosting areas and wind farms. Conflicts between wind farms and the occurrence of roosting birds are expected to be common, particularly when wind farms are further extended inland. This also applies in general terms to bats, for which so far almost nothing is known about how wind farms restrict their hunting areas.

So far, an overarching study to analyse what specific habitat parts of particularly sensitive species (e.g. geese) have been affected by wind farms is also missing. Such studies should be carried out, if possible, in Europe, definitely throughout Germany, and should contain an analysis of the distribution pattern of wind farms and non-breeding birds, possibly also of breeding birds.

Furthermore long-term studies are necessary to analyse the long-term effects of wind farms, which so far are almost completely unknown. These effects could, for example include „habituation" of birds and bats to wind farms, so that in the long term, negative effects would become less so. Such a phenomenon is not expected for migrating birds or bats, but nonetheless it cannot be excluded that population declines are only detectable over the longer term. This could be the case for long-established breeding birds (e.g. relatively long-living meadow birds), which because of their distinct site-faithfulness are badly affected by disturbance displacement due to wind farms, and consequently their abandoned territories are not occupied by future generations.

The possible impacts of wind farms as barriers to migrating birds, are still too little known to be able to draw any conclusions. It is not known under what circumstances migration is impeded, nor is it possible to determine if the potential negative effects are relevant to the course of migration. Here, in addition to the traditional visual observations the use of specialised equipment at night is required. Spatial analysis of wind farm sites is also needed, in order to calculate the probability of birds colliding with wind farms on particular migration routes in Europe.

Even though current relationships between wind turbine height and displacement effects give preliminary indications of the likely effects of larger wind turbines in the context both of repowering and of further expansion of wind energy production, the results are on their own can hardly be considered robust. Studies examining the displacement effects of different large wind farms -if possible- under the same conditions with identical methodologies, are urgently needed. They should be carried out over a period of at least two years in an area where there are many wind turbines and where lots of roosting birds are expected. Studies investigating the disturbance effects of replacement wind turbines are required for at least two years before and after repowering is carried out.

Surprisingly, only a few studies exist covering the reactions towards wind farms of many sensitive bird species (e.g. storks, birds of prey, crane, but also corncrake (Müller & Illner, 2002) and therefore species-specific studies are urgently needed.

In Germany, almost no systematic analyses of the numbers of bird and bat casualties are available. The data compiled by T. Dürr (Tab. 9 and 11) indicate that only relatively few bird species are faced with large problems (birds of prey, gulls and certain bat species), but cannot reveal much about the actual scale of the losses. Because corpse searching is undertaken with varying effort, sometimes only for very short periods, true numbers of collision victims remain unknown. The effects of the new generation of very large wind turbines with their warning lights on nocturnal migrant birds are also totally unknown and hence, systematic analyses are urgently required of bird and bat mortality at very large wind turbines at potential danger points (well-known migration hotspots). In addition, the methodical guidelines from USA should be applied and adapted to German conditions (Anderson et al., 2000a; Morrison, 1998; Smallwood & Thelander, 2004). In any case, studies must determine experimentally the disappearance rate of corpses and take into account the search efficiency of the researchers at the different sites.

Good prospects [for research] would be the development of methods and equipment for automatic recording of collisions of birds and bats (thermal-imaging cameras, radar equipment, see also Cooper & Kelly, 2000; Desholm, 2003; Erickson et al., 2001; Verhoef et al., 2002). In this way, collisions are recorded more efficiently in the long run and at difficult sites actually for the first time (e.g. on the open sea). It is also possible to find out more about the circumstances in which birds and bats are killed and in the light of knowledge, to react to particular situations of high risk (temporarily switching off turbines during periods of heavy bird migration or bad weather). The development and testing of automatic recording equipment should be pursued as a matter of priority. It is also necessary to determine the weather conditions, which present particular risks to birds and bats.

Very little is known about the impact of additional mortality due to wind turbines on the population dynamics of the species of bird and bat, which are affected. The results presented in chapter 4 only scratch the surface of the problem. Rates of collision mortality due to wind turbines could be estimated from numbers of known casualties of the species most greatly affected. Then, population models have to take into account the ability of the population to recover. In particular, there is a need to establish the extent to which the population is able to balance any losses by (density-dependent) increases in productivity, and to analyse all signs of density-dependence in reproductive and mortality rates.

A particular problem of wind energy in Germany is the high collision rate of red kites. Mostly local breeding birds are involved and in many cases the brood might [also] be lost following a collision involving a breeding adult. Because of the high responsibility Germany bears for this species (around half of the world population of red kites breeds in Germany) and because the world population is small (24.000 pairs BirdLife International & European Bird Census Council, 2000), steps to solve this problem have to be taken urgently. The following measures seem to be necessary:

1. Analysis of the circumstances in which casualties listed to date in the register of the „Staatl. Vogelschutzwarte Brandenburg" were found;
2. Monitoring of all main areas where red kites occur in Germany must be established, in order to register the spatial and temporal scale of this phenomenon. Because of the size and greater chances of finding these birds, as well as the fact that they are only present in the breeding season (April to Au-

gust), the methods should be less demanding than general collision monitoring;
3. Analysis of hunting behaviour and the use of habitat by red kites close to wind farms where already losses have been recorded (observations of behaviour and fitting birds with radio transmitters);
4. Provided that steps 1-3 give concrete results, experimental development of habitat management techniques at wind farms (measures to deter birds, reduce food supply etc.) and verification of their effectiveness.

A similar procedure should be considered for white-tailed eagles.

The possibilities referred to in chapter 5 to reduce collision rates of birds, have so far not been tested in the field, and therefore field trials are essential (Erickson et al., 2001).

Because of the collisions of bats with wind turbines, there is a need for basic neuro-physiological or behavioural studies, with the aim of clarifying the orientation mechanism of bats when approaching wind turbines. One possibility would be to develop measures, which force bats to use their sonar-orientation.

In the case of corncrake, quail and possibly also other species, behavioural experiments should clarify whether noises generated by wind turbines prevents the acoustic communication by these species.

Research regarding other forms of renewable energy production

The need for research into the other forms of renewable energy production examined here is even higher than for wind energy. This is particularly true for solar energy parks. The effects of solar power stations on breeding and migrating birds, as well as on other animals and plants in their environment are not known. Nor is any data available on the disturbance displacement, or on collision mortality, due to water birds mistaking solar cells for water surfaces at night. Studies analysing the extent of impacts of solar parks on bird populations (and on other taxa) are urgently needed. These studies should be carried out in the similar way as for wind farms. The populations must be observed at least over two years before and at least two years after the installation of the solar power plant. In addition, a control site should be set up which is similar to the solar power plant, but without solar cells. As well as mapping the numbers and distribution of breeding and non-breeding birds, the area should also be searched for collision victims. The same criteria for the equivalent studies at wind farms (see above) apply here too. In particular, the disappearance of corpses needs to be controlled. Only limited data are available for the effects of energy crops, though in some cases knowledge of agricultural impacts on birds can help.

Table 19: Most important topics for future research

Form of energy production	Research topic	Priority
Wind farm	Disturbance effect dependent on wind farm size	high
Wind farm	Collision rates of birds and bats (per year) dependent on wind farm size	very high
Wind farm	Development and testing of equipment to record collisions automatically	high
Wind farm	Special study on red kites	very high
Wind farm	Special study on sensitive birds of prey, corncrakes and other species	high
Wind farm	Before-after studies of bird and bat populations	-
Wind farm	Long-term effects of wind farms	-
Wind farm	Influence of the availability of habitats on disturbance displacement	-
Wind farm	Widespread disturbance of bird populations due to wind farms (what extent of land with no wind farms is still available in Germany/Europe?)	-
Wind farm	Barrier effects of wind farms	-
Wind farm	Precise population models to assess collision mortality	-
Wind farm	Study of the orientation methods of migrating bats	high
Wind farm	Study of the influence of wind farms on acoustic communication between specific bird species	-
Solar-park	Before-after studies of breeding and non-breeding bird populations	very high
Solar-park	Collision rates at solar parks	very high
Bio-mass	Impacts of extensive bio-mass cultivation on biodiversity	high

9 Acknowledgements

For supporting our work with discussions and encouragement, for courteously providing data and literature as well as other help, we would like to thank: Kathrin Ammermann, Yannick André, Karin Andrick, Lothar Bach, Tobias Dürr, Klaus-Michael Exo, Bernd Hälterlein, Friedhelm Igel, Hubertus Illner, Bettina Keite, Claus Mayr, Frank Musiol, Hans-Ulrich Rösner, Matthias Schreiber, Anna Ziege and all other members of the project-accompanying work-team to this project. The statistical analysis was supported by Professor Les G. Underhill, Avian Demography Unit, Department of Statistical Sciences, University of Cape Town. We thank Solveigh Lass-Evans for translating the text.

10 Literature

ACHA, A. (1998). Negative impact of wind generators on Eurasion Griffon Gyps fulvus in Tarifa, Spain. Vulture News 38, 10-18.

AG EINGRIFFSREGELUNG. (1996). Empfehlungen zur Berücksichtigung der Belange des Naturschutzes und der Landschaftspflege beim Ausbau der Windenergienutzung. Natur und Landschaft 71, 381-385.

AHLÉN, I. (2002). Fladdermöss och fåglar dödade av vindkraftverk. Fauna och Flora 97, 14-21.

ALBOUY, S., CLÉMENT, D., JONARD, A., MASSÉ, P., PAGÈS, J.-M. & NEAU, P. (1997). Suivi ornithologique du parc éolien de Porte-la-Nouvelle (Aude) - Rapport final. ABIES, LPO, Gardouch.

ALBOUY, S., DUBOIS, Y. & PICQ, H. (2001). Suivi ornithologique des parcs éoliens du plateau de Garrigue Haute (Aude) - Rapport final. ABIES, LPO, Gardouch.

ANDERSON, G. Q. A., HASKINS, L. R. & NELSON, S. H. (2003). The effects of bioenergy crops on farmland birds in the UK - a review of current knowledge and future predictions. RSPB, Sandy.

ANDERSON, R., MORRISON, M., SINCLAIR, K. & STRICKLAND, D. (1999). Studying Wind Energy/Bird Interactions: A Guidance Document. Avian Subcommittee and the National Wind Coordinating Committee, Washington, DC.

ANDERSON, R. L., MORRISON, M. L., SINCLAIR, K. & STRICKLAND, M. D. (2000a). Studying Wind Energy/Bird Interactions: A Guidance Document - Executive Summary. In Proceedings of National Avian - Wind Power Planning Meeting III (ed. PNAWPPM-III), pp. 126-131. Prepared for the Avian Subcommittee of the National Wind Coordinating Committee by LGL Ltd., King City, Ont., San Diego, California.

ANDERSON, R. L., STRICKLAND, M. D., TOM, J., NEUMANN, N., ERICKSON, W. P., CLECKLER, J., MAYORGA, G., NUHN, G., LEUDERS, A., SCHNEIDER, J., BACKUS, L., BECKER, P. & FLAGG, N. (2000b). Avian Monotoring and Risk Assessment at Tehachapi Pass and San Gorgonio Pass Wind Resource Areas, California. In Proceedings of National Avian - Wind Power Planning Meeting III (ed. PNAWPPM-III), pp. 31-46. Prepared for the Avian Subcommittee of the National Wind Coordinating Committee by LGL Ltd., King City, Ont., San Diego, California.

BACH, L. (2001). Fledermäuse und Windenergienutzung - reale Probleme oder Einbildung? Vogelkundliche Berichte aus Niedersachsen 33, 119-124.

BACH, L. (2002). Auswirkungen von Windenergieanlagen auf das Verhalten von Fledermäusen am Beispiel des Windparks „Hohe Geest", Midlum. Bericht der Arbeitsgemeinschaft zur Förderung angewandter biologischer Forschung im Auftrag der KW Midlum GmbH & Co. KG, Freiburg, Niederelbe.

BACH, L. (2003). Effekte von Windenergieanlagen auf Fledermäuse. In: Kommen die Vögel und Fledermäuse unter die (Wind)räder?, Dresden, 17.-18.11.2003.

BACH, L., HANDKE, K. & SINNING, F. (1999). Einfluss von Windenergieanlagen auf die Verteilung von Brut- und Rastvögeln in Nordwest-Deutschland. Bremer Beiträge für Naturkunde und Naturschutz 4, 107-122.

BARRIOS, L. & RODRIGUEZ, A. (2004). Behavioural and environmental correlates of soaring bird mortality at on-shore wind turbines. Journal of Applied Ecology 41, 72-81.

BEINTEMA, A.J. & MÜSKENS, G.J.D.M. (1981), De invloed van beheer op de productiviteit van weidevogels. RIN-rapport 81/19, Leersum.

BEINTEMA, A. J. & DROST, N. (1986). Migration of the Black-tailed Godwit. Le Gerfaut 76, 37-62.

BEZZEL, E. (1985), Kompendium der Vögel Mitteleuropas. Aula, Wiesbaden.

BEZZEL, E. (1993), Kompendium der Vögel Mitteleuropas. Aula, Wiesbaden.

BERGEN, F. (2001a). Untersuchungen zum Einfluss der Errichtung und des Betriebs von Windenergieanlagen auf Vögel im Binnenland. Dissertation, Ruhr Universität Bochum.

BERGEN, F. (2001b). Windkraftanlagen und Frühjahrsdurchzug des Kiebitz (Vanellus vanellus): eine Vorher/Nacher-Studie an einem traditionellen Rastplatz in Nordrhein-Westfalen. Vogelkundliche Berichte aus Niedersachsen 33, 89-96.

BERGEN, F. (2002a). Einfluss von Windenergieanlagen auf die Raum-Zeitnutzung von Greifvögeln. In Windenergie und Vögel - Ausmaß und Bewältigung eines Konfliktes (ed. H. Ohlenburg), pp. 86-96. Technische Universität, Berlin.

BERGEN, F. (2002b). Windkraftanlagen und Frühjahrsdurchzug des Kiebitz (Vanellus vanellus): eine Vorher-Nachner-Studie an einem traditionellen Rastplatz in Nordrhein-Westfalen. In Windenergie und Vögel - Ausmaß und Bewältigung eines Konfliktes (ed. H. Ohlenburg), pp. 77-85. Technische Universität, Berlin.

BERGH, L. M. J. v. D., SPAANS, A. L. & SWELM, N. D. v. (2002). Lijnopstellingen van windturbines geen barrière voor voedselvluchten van meeuwens en sterns in de broedtijd. Limosa 75, 25-32.

BEYEA, J., COOK, J. H. & HOFFMAN, W. (1995). Vertebrate species diversity in large-scale energy crops and associated policy issues. Annual progress report to Biofuels Feedstock Development Program. Oak Ridge National Laboratory, Oak Ridge.

BIBER, O. (1993). Raumnutzung der Goldammer Emberiza citrinella für die Nahrungssuche zur Brutzeit in einer intensiv genutzten Agrarlandschaft (Schweizer Mittelland). Ornithologischer Beobachter 90, 283-296.

BIOCENOSE & LPO AVEYRON - GRANDS CAUSSES. (2002). Synthèse et analyse bibliographique visant à évaluer les impacts des éoliennes sur les populations de vertébrés sauvages. BIOCENOSE et LPO Aveyron - Grands Causses pou ADEME, Onet-le-Chateau.

BIRDLIFE INTERNATIONAL & EUROPEAN BIRD CENSUS COUNCIL. (2000). European bird populations: estimates and trends. BirdLife International, Cambridge.

BMU. (2004). Themenpapier Windenergie. Bundesministerium für Umwelt, Naturschutz und Reaktorsicherheit. Art 2122, März 2004.

BOONE, D. (2003). Bat kill at West Virginia windplant, Maryland.

BÖTTGER, M., CLEMENS, T., GROTE, G., HARTMANN, G., HARTWIG, E., LAMMEN, C., VAUK-HENTZELT, E. & VAUK, G. (1990). Biologisch-Ökologische Begleituntersuchungen zum Bau und Betrieb von Windkraftanalgen. In NNA-Berichte 3, Sonderheft. NNA.

BRAUNEIS, W. (1999). Der Einfluss von Windkraftanlagen auf die Avifauna am Beispiel der „Solzer Höhe" bei Bebra-Solz im Landkreis Hersfeld-Rotenburg - Untersuchung im Auftrag des Bundes für Umwelt und Naturschutz (BUND) Landesverband Hessen e. V. - Ortsverband Alheim-Rotenburg-Bebra, pp. 1-91, Bebra.

BRAUNEIS, W. (2000). Der Einfluss von Windkraftanlagen (WKA) auf die Avifuna, dargestellt insb. am Beispiel des Kranichs Grus grus. Ornithologische Mitteilungen 52, 410-415.

BUNGE, T., DIRBACH, D., DREHER, B., FRITZ, K., LELL, O., RECHENBERG, B., RECHENBERG, J., SCHMITZ, E., SCHWERMER, S., STEINHAUER, M., STEUDTE, C. & VOIGT, T. (2001), Wasserkraftanlagen als erneuerbare Energiequelle - rechtliche und ökologische Aspekte. Umweltbundesamt.

BURNHAUSER, A. (1983). Zur ökologischen Situation des Weißstorchs in Bayern: Brutbestand, Biotopansprüche, Schutz und Möglichkeiten der Bestandserhaltung und -verbesserung. Abschlußbericht, Inst. f. Vogelk., Garm.-Partenk., 488 S.

BURTON, N. H. K., WATTS, N. P., CRICK, H. Q. P. & EDWARDS, P. J. (1999). The effect of preharvesting operations on Reed Buntings Emberiza schoeniclus nesting in Oilseed Rape Brassica napus. Bird Study 46, 369-372.

CATCHPOLE, E. A., MORGAN, B. J. T., FREEMAN, S. N. & PEACH, W. J. (1999). Modelling the survival of British Lapwings Vanellus vanellus using ring-recovery data and weather covariates. Bird Study 46 (supplement), 5-13.

CLEMENS, T. & LAMMEN, C. (1995). Windkraftanlagen und Rastplätze von Küstenvögeln - ein Nutzungskonflikt. Seevögel 16, 34-38.

COOPER, B. A. & KELLY, T. A. (2000). Night Vision and Thermal Imaging Equipment. In Proceedings of National Avian - Wind Power Planning Meeting III (ed. PNAWPPM-III), pp. 164-165. Prepared for the Avian Subcommittee of the National Wind Coordinating Committee by LGL Ltd., King City, Ont., San Diego, California.

CRAWFORD, R. L. & ENGSTROM, R. T. (2001). Characteristics of avian mortality at a north Florida television tower: a 29-year study. Journal of Field Ornithology 72, 380-388.

CROCKFORD, N. J. (1992). A review of the possible impacts of wind farms on birds and other wildlife. In JNCC Report, vol. 27, pp. 60, Peterborough.

DE LUCAS, M., JANSS, G. F. E. & FERRER, M. (2004). The effects of a wind farm on birds in a migration point: the Strait of Gibraltar. Biodiversity and Conservation 13, 395-407.

DESHOLM, M. (2003). Thermal animal detection systems (TADS). Development of a method for estimating collision frequency of migrating birds at offshore wind turbines. NERI Technical Report No. 440.

DIERSCHKE, V., HÜPPOP, O. & GARTHE, S. (2003). Populationsbiologische Schwellen der Unzulässigkeit für Beeinträchtigungen der Meeresumwelt am Beispiel der in der deutschen Nord- und Ostsee vorkommenden Vogelarten. Seevögel 24, 61-72.

DONALD, P. F., EVANS, A. D., BUCKINGHAM, D. L., MUIRHEAD, L. B. & WILSON, J. D. (2001). Factors affecting the territory distribution of Skylarks Alauda arvensis breeding on lowland farmland. Bird Study 48, 271-278.

DONALD, P. F., EVANS, A. D., MUIRHEAD, L. B., BUCKINGHAM, D. L., KIRBY, W. B. & SCHMITT, S. I. A. (2002). Survival rates, causes of failure and productivity of Skylark Alauda arvensis nests on lowland farmland. Ibis 144, 652-664.

DOOLING, R. J. & LOHR, B. (2001). The Role of Hearing in Avian Avoidance of Wind Turbines. In Proceedings of National Avian - Wind Power Planning Meeting IV (ed. PNAWPPM-IV), pp. 115-127. Prepared for the Avian Subcommittee of the National Wind Coordinating Committee by RESOLVE, Inc., Washington, D.C., Susan Savitt Schwartz, Carmel, California.

DÖRING, V. & HELFRICH, R. (1986). Zur Ökologie einer Rebhuhnpopulation (Perdix perdix, Linné, 1758) im Unteren Naheland (Rheinland-Pfalz, Bundesrepublik Deutschland). Enke, Stuttgart.

DULAS ENGINEERING LTD. (1995). The Mynydd y Cemmaes windfarm impact study. Vol. IID - Ecological impact - final report. ETSU report: W/13/00300/REP2D.

DÜRR, T. (2001). Verluste von Vögeln und Fledermäusen durch Windkraftanlagen in Brandenburg. Otis 9, 123-125.

DÜRR, T. (2003a). Neue Seeadler-Verluste an Windenergieanlagen in Deutschland. In: Projektgruppe Seeadlerschutz, Jahresbericht 2003, Kiel.

DÜRR, T. (2003b). Windenergieanlagen und Fledermausschutz in Brandenburg - Erfahrungen aus Brandenburg mit Einblick in die bundesweite Fundkartei von Windkraftopfern. In: Kommen die Vögel und Fledermäuse unter die (Wind)räder?, Dresden, 17.-18.11.2003.

DÜRR, T. (2004). Vögel als Anflugopfer an Windenergieanlagen - ein Einblick in die bundesweite Fundkartei. Bremer Beiträge für Naturkunde und Naturschutz im Druck.

DÜRR, T. & BACH, L. (2004). Fledermäuse als Schlagopfer von Windenergieanlagen - Stand der Erfahrungen mit Einblick in die bundesweite Fundkartei. Bremer Beiträge für Naturkunde und Naturschutz im Druck.

DURELL, S. E. A. L. V. D. & GOSS-CUSTARD, J. D. (2000). Density-dependent mortality in Oystercatchers Haematopus ostalegus. Ibis 142, 132-138.

EAS. (1997). Ovenden Moor Ornithological Monitoring. Report to Yorkshire Windpower. Keighly: Ecological Advisory Service.

EBBINGE, B. S., VAN BIEZEN, J. B. & VAN DER VOET, H. (1991). Estimation of annual adult survival rates of Barnacle Geese *Branta leucopsis* using multiple resightings of marked individuals. Ardea 79, 73-112.

ERICKSON, W., JOHNSON, G., YOUNG, D., STRICKLAND, D., GOOD, R., BOURASSA, M., BAY, K. & SERNKA, K. J. (2002). Synthesis and comparison of baseline avian and bat use, raptor nesting and mortality information from proposed and existing wind developments, pp. 1-60. Report for Bonneville Power Administration, Portland, Oregon.

ERICKSON, W., KRONNER, K. & GRITSKI, B. (2003). Nine Canyon Wind Power Project. Avian and Bat Monitoring Report. September 2002 - August 2003. Prepared for Nine Canyon Technical Advisory Committee by West, Inc., Cheyenne.

ERICKSON, W. P., JOHNSON, G. D., STRICKLAND, M. D., KRONNER, K., BECKER, P. S. & ORLOFF, S. (1999). Baseline avian use and behavior at the Cares Wind Plant Site, Klickitat County, Washington. NREL/SR-500-26902.

ERICKSON, W. P., JOHNSON, G. D., STRICKLAND, M. D., YOUNG, D. P., JR., SERNKA, K. J. & GOOD, R. E. (2001). Avian collisions with wind turbines: a summary of existing studies and comparison to other sources of avian collision mortality in the United States National Wind Coordinating Comitee (NWCC). Western EcoSystems Technology Inc., Washington D.C.

EVERAERT, J. (2003). Collision victims on 3 wind farms in Flanders (Belgium) in 2002. Instituut voor Naturbeheer, Brussel.

EVERAERT, J., DEVOS, K. & KUIJKEN, E. (2002). Windturbines en vogels in Vlaanderen. Instituut voor Natuurbehoud, Brussels.

EVERAERT, J., DEVOS, K. & KUIJKEN, E. (2003). Vogelconcentraties en vliegbewegingen in Vlaanderen. Instituut voor Natuurbehoud, Brussels.

FERNANDEZ-DUQUE, E. & VALEGGIA, C. (1994). Meta-analysis: a valuable tool in conservation research. Conservation Biology 8, 555-561.

FÖRSTER, F. (2003). Windkraft und Fledermausschutz in der Oberlausitz. In Kommen die Vögel und Fledermäuse unter die (Wind)räder?, Dresden, 17.-18.11.2003.

FUCHS, S. (1997). Nahrungsökologie handaufgezogener Rebhuhnküken - Effekte unterschiedlicher Formen und Intensitäten der Landnutzung, Diplomarbeit, Freie Universität Berlin.

GANTER, B., LARSSON, K., SYROECHKOVSKY, E. V., LITVIN, K. E., LEITO, A. & MADSEN, J. (1999), Barnacle Goose Branta leucopsis: Russia/Baltic. In: MADSEN, J., CRACKNELL, G. & FOX, A. D. (Hrsg): Barnacle Goose Branta leucopsis: Russia/Baltic. Wetlands International Publ. No. 48, National Environmental Research Institute, Wageningen and Rönde.

GERJETS, D. (1999). Annäherung wiesenbrütender Vögel an Windkraftanalagen - Ergebnisse einer Brutvogeluntersuchung im Nahbereich des Windparks Drochtersen. Bremer Beiträge für Naturkunde und Naturschutz 4, 49-52.

GHARADJEDAGHI, B. & EHRLINGER, M. (2001). Auswirkungen des Windparks bei Nitzschka (Lkr. Altenburger Land) auf die Vogelfauna. Landschaftspflege und Naturschutz in Thüringen 38, 73-83.

GILLINGS, S. (2001). Factors affecting the distribution of skylarks Alauda arvensis wintering in Britain and Ireland during the early 1980s. In The ecology and conservation of skylarks Alauda arvensis (ed. P. F. Donald and J. A. Vickery), pp. 115-128. RSPB, Sandy.

GLUTZ VON BLOTZHEIM, U. N. & BAUER, K. M. (1997), Handbuch der Vögel Mitteleuropas. Band 14. Passeriformes (5.Teil). AULA, Wiesbaden.

GORANSEN, G. (1990). Energy foresting in agricultural areas and changes in the avifauna. Norv. Ser. C, Cinclus Suppl. 1, 17-20.

GRANT, M. C., LODGE, C., MOORE, N., EASTON, J., ORSMAN, C., SMITH, M., THOMPSON, G. & RODWELL, S. (1999). Breeding success and causes of breeding failure of curlew Numenius arquata in Northern Ireland. Journal of Applied Ecology 36, 59-74.

GREEN, R. E. (1999). Survival and dispersal of male Corncrakes Crex crex in a threatened population. Bird Study 46 (supplement), 218-229.

GROEN, N. M. & HEMERIK, L. (2002). Reproductive success and survival of Black-tailed Godwits Limosa limosa in a declining local population in the Netherlands. Ardea 90, 239-248.

GUILLEMETTE, M. & LARSEN, J. K. (2002). Postdevelopment experiments to detect anthropogenic disturbances: the case of sea ducks and wind parks. Ecological Applications 12, 868-877.

GUILLEMETTE, M., LARSEN, J. K. & CLAUSANGER, I. (1999). Assessing the impact of the Tunø Knob wind park on sea ducks: the influence of the food ressources. National Environmental Research Institute, Denmark.

HALL, L. S. & RICHARDS, G. C. (1962). Notes on Tadarida australis (Chiroptera: molossidae). Australian Mammology 1, 46.

HODOS, W. (2001). Minimization of Motion Smear: Reducing Avian Collisions with Wind Turbines. NREL/SR-500-33249, Maryland.

HODOS, W., POTOCKI, A., STROM, T. & GAFFNEY, M. (2001). Reduction of Motion Smear to Reduce Avian Collisions with Wind Turbines. In Proceedings of National Avian - Wind Power Planning Meeting IV (ed. PNAWPPM-IV), pp. 88-106. Prepared for the Avian Subcommittee of the National Wind Coordinating Committee by RESOLVE, Inc., Washington, D.C., Susan Savitt Schwartz, Carmel, California.

HÖTKER, H. (1990), Der Wiesenpieper. Ziemsen, Wittenberg.

HOOVER, S. (2002). The response of Red-tailed Hawks and Golden Eagles to topographical features, weather, and abundance of a dominant prey species at the Altamont Pass Wind Ressource Area, California. NREL/SR-500-30868.

HORMANN, M. (2000). Schwarzstorch - Ciconia nigra. In Avifauna von Hessen, 4. Lieferung (ed. H. G. f. O. u. Naturschutz).

HUNT, G. (2002). Golden Eagles in a perilous landscape: predicting the effects of mitigation for wind turbine blade-strike mortality. Consultation Report to California Energy Commission.

HYDRO TASMANIA. Bird and bat monitoring. Hydro Tasmania.

IKSR, CIPR & ICBR (2004). Ökologische Auswirkung von (Klein)-Wasserkraftanlagen auf die Lebensbedingungen von Wanderfischen. 1-14.

IMMELMANN, K. (1976), Einführung in die Verhaltenskunde. Parey, Berlin.

INGLIS, I. R., ISAACSON, A. J., SMITH, G. C., HAYES, P. J. & THEARLE, R. J. P. (1997). The effect on the woodpigeon (Columba palumbus) of the introduction of oilseed rape into Britain. Agriculture, Ecosystems and Environment 61, 113-121.

ISSELBÄCHER, K. & ISSELBÄCHER, T. (2001). Vogelschutz und Windenergie in Rheinland-Pfalz. In Naturschutz und Landschaftspflege, pp. 1-183, Oppenheim.

JANSS, G. (2000). Bird Behaviour In and Near a Wind Farm at Tarifa, Spain: Management Considerations. In Proceedings of National Avian - Wind Power Planning Meeting III (ed. PNAWPPM-III), pp. 110-114. Prepared for the Avian Subcommittee of the National Wind Coordinating Committee by LGL Ltd., King City, Ont., San Diego, California.

JENNY, M., WEIBEL, U., LUGRIN, B., JOSEPHY, B., REGAMEY, J.-L. & ZBINDEN, N. (2002). Rebhuhn. Schlussbericht 1991-2000. Bundesamt für Umwelt, Wald und Landschaft in Zusammenarbeit mit der Schweizerische Vogelwarte, Bern.

JOHNSON, G. D. (2002). What is known and not known about impacts on bats? In Proceedings of the Avian Interactions with Wind Power Structures, October 16-17, 2002 (in press), Jackson Hole, Wyoming.

JOHNSON, G. D., ERICKSON, W. P., STRICKLAND, D. M., SHEPHERD, M. F., SHEPHERD, D. A. & SARAPPO, S. A. (2003). Mortality of Bats at a Large-scale Wind Power Development at Buffalo Ridge, Minnesota. Am. Midl. Nat. 150, 332-342.

JOHNSON, G. D., YOUNG, D. P., ERICKSON, W. P., DERBY, C. E., STRICKLAND, M. D. & GOOD, R. E. (2000). Wildlife monitoring studies Sea West Windpower Project, Carbon County, Wyoming. Western EcoSystems Technology, Inc., Cheyenne.

KAATZ, J. (2000). Untersuchungen zur Avifauna im Bereich des Windparks Badeleben im Bördekreis - Standort- und zeitbezogene Habitatnutzung von Brut- und Rastvögeln im Prä-Post-Test-Verfahren, pp. 1-38. IHU Geologie und Analytik, Neuruppin.

KAATZ, J. (2002). Artenzusammensetzung und Dominanzverhältnisse einer Heckenbütergemeinschaft im Windfeld Nackel. In Windenergie und Vögel - Ausmaß und Bewältigung eines Konfliktes (ed. H. Ohlenburg), pp. 113-124. Technische Universität, Berlin.

KARLSSON, J. (1983). Birds and windpower, pp. 12.

KAVANAGH, B. (1990). Bird communities of two short rotation forestry plantations on cutover peatland. Irish Birds 4, 169-180.

KEELEY, B., UGORETZ, S. & STRICKLAND, M. D. (2001). Bat Ecology and Wind Turbine Considerations. In Proceedings of National Avian - Wind Power Planning Meeting IV (ed. PNAWPPM-IV), pp. 135-146. Prepared for the Avian Subcommittee of the National Wind Coordinating Committee by RESOLVE, Inc., Washington, D.C., Susan Savitt Schwartz, Carmel, California.

KELLY, T. A. (2000). Radar, Remote Sensing and Risk Management. In Proceedings of National Avian - Wind Power Planning Meeting III (ed. PNAWPPM-III), pp. 152-161. Prepared for the Avian Subcommittee of the National Wind Coordinating Committee by LGL Ltd., King City, Ont., San Diego, California.

KERLINGER, P. (2000). An Assessment of the Impacts of Green Mountain Power Corporation's Searsburg, Vermont, Wind Power Facility on Breeding and Migrating Birds. In Proceedings of National Avian - Wind Power Planning Meeting III (ed. PNAWPPM-III), pp. 90-96. Prepared for the Avian Subcommittee of the National Wind Coordinating Committee by LGL Ltd., King City, Ont., San Diego, California.

KETZENBERG, C., EXO, K.-M., REICHENBACH, M. & CASTOR, M. (2002). Einfluss von Windenergieanlagen auf brütende Wiesenvögel. Natur und Landschaft 77, 144-153.

KOKS, B. J., SCHARENBURG, C. W. M. V. & VISSER, E. G. (2001). Grauwe Kiekendieven Circus pygargus in Nederland: balanceren tussen hoop en vrees. Limosa 74, 121-136.

KOOP, B. (1997). Vogelzug und Windenergieplanung. Beispiele für Auswirkungen aus dem Kreis Plön (Schlewsig-Holstein). Naturschutz und Landschaftsplanung 29, 202-207.

KOOP, B. (1999). Windkraftanlagen und Vogelzug im Kreis Plön. Bremer Beiträge für Naturkunde und Naturschutz 4, 25-32.

KORN, M. & SCHERNER, R. (2000). Raumnutzung von Feldlerchen (Alauda arvensis) in einem Windpark. Natur und Landschaft 75, 74-75.

KOSTREWA, A. & G. SPEER (HRSG.) (1995). Greifvögel in Deutschland. Aula-Verl., Wiesbaden, 113 S.

KOWALLIK, C. & BORBACH-JAENE, J. (2001). Windräder als Vogelscheuchen? - Über den Einfluss der Windkraftnutzung in Gänserastgebieten an der nordwestdeutchen Küste. Vogelkundliche Berichte aus Niedersachsen 33, 97-102.

KRAPP, F. (HRSG.) (2001). Handbuch der Säugetiere Europas, Band 4: Fledertiere, Teil I: Chiroptera 1. Aula-Verl., Wiesbaden, 602 S.

KRAPP, F. (HRSG.) (2004). Handbuch der Säugetiere Europas, Band 4: Fledertiere, Teil II: Chiroptera II,. Aula-Verl., Wiesbaden, S. 606 – 1186.

KRUCKENBERG, H. & BORBACH-JAENE, J. (2001). Auswirkungen eines Windparks auf die Raumnutzung nahrungssuchender Blessgänse - Ergebnisse aus einem Monitoringprojekt mit Hinweisen auf ökoethologischen Forschungsbedarf. Vogelkundliche Berichte aus Niedersachsen 33, 103-109.

KRUCKENBERG, H. & JAENE, J. (1999). Zum Einfluss eines Windparks auf die Verteilung weidender Bläßgänse im Rheiderland (Landkreis Leer, Niedersachsen). Natur und Landschaft 74, 420-427.

KUTSCHER, J. (2002). Ökologische Begleitforschung zur Offshore-Windenergienutzung. Fachtagung des Bundesministeriums für Umwelt, Naturschutz und Reaktorsicherheit und des Projektträgers Jülich, Bremerhaven.

LANGSTON, R. (2002). Wind Energy and Birds: Results and Requirements. In RSPB Research Report No. 2, pp. 1-54. RSPB, Sandy.

LANGSTON, R. W. H. & PIULLAN, J. D. (2003). Wind farms and birds: an analysis of the effects of wind farms on birds, and guidance on environmental assessment criteria and site selection issues. Report written by BirdLife International on behalf of the Bern Convention, Sandy.

LEDDY, K. L., HIGGINS, K. F. & NAUGLE, D. E. (1999). Effects of wind turbines on upland nesting birds in Conservation Reserve Program grasslands. Wilson Bulletin 111, 100-104.

LEKUONA, J. M. (2001). Uso del espacio por la avifauna y control de la mortalidad de aves y murciélagos en los parques eólicos de Navarra durante un ciclo anual. Direccion General de Medio Ambiente, Gobierno de Navarra, Pamplona.

MANVILLE, A. M. (2001). Communication Towers, Wind Generators, and Research: Avian Conservation Concerns. In Proceedings of National Avian - Wind Power Planning Meeting IV (ed. PNAWPPM-IV), pp. 152-159. Prepared for the Avian Subcommittee of the National Wind Coordinating Committee by RESOLVE, Inc., Washington, D.C., Susan Savitt Schwartz, Carmel, California.

MARQUENIE, J.M. & VAN DE LAAR, F. (2004). Impacts on Biodiversity: Offshore drilling and production platforms and bird migration. Manuskript.

MASON, C. F. & MACDONALD, S. M. (2000). Influence of landscape and land-use on the distribution of breeding birds in farmland in eastern England. Journal of Zoology 251, 338-348.

McISAAC, H. P. (2001). Raptor Acuity and Wind Turbine Blade Conspicuity. In Proceedings of National Avian - Wind Power Planning Meeting IV (ed. PNAWPPM-IV), pp. 59-87. Prepared for the Avian Subcommittee of the National Wind Coordinating Committee by RESOLVE, Inc., Washington, D.C., Susan Savitt Schwartz, Carmel, California.

MEEK, E. R., RIBBANDS, J. B., CHRISTER, W. G., DAVEY, P. R. & HIGGINSON, I. (1993). The effects of aero-generators on moorland bird populations in the Orkney Islands, Scotland. Bird Study 40, 140-143.

MENZEL, C. (2002). Rebhuhn und Rabenkrähe im Bereich von Windkraftanlagen im niedersächsischen Binnenland. In Windenergie und Vögel - Ausmaß und Bewältigung eines Konfliktes (ed. H. Ohlenburg), pp. 97-112. Technische Universität, Berlin.

MENZEL, C. & POHLMEIER, K. (1999). Indirekter Raumnutzungsnachweis verschiedener Niederwildarten mit Hilfe von Losungsstangen („dropping marker") in Gebieten mit Winkraftanlagen. Z. Jagdwiss. 45, 223-229.

MEYERHOFF, J., PETSCHOW, U., HERRMANN, N., KEHREN, M. & LEIFELD, D. (1998), Umweltverträglichkeit kleiner Wasserkraftwerke - Zielkonflikte zwischen Klima- und Gewässerschutz. Umweltbundesamt.

MÖLLER, B. & A. NORTTORF (1997). Der Schwarzstorch (Ciconia nigra) in Niedersachsen - Aktuelle und historische Bestandssituation, Reproduktion, Habitatansprüche und Schutzmaßnahmen -. Vogelkdl. Ber. Niedersachs. 29: 51-61.

MOOIJ, J. H., FARAGÓ, S. & KIRBY, J. S. (1999), White-fronted Goose *Anser albifrons albifrons*. In: MADSEN, J., CRACKNELL, G. & FOX, A. D. (Hrsg): White-fronted Goose *Anser albifrons albifrons*. Wetlands International Publ. No. 48, National Environmental Research Institute, Wageningen and Rönde.

MORRIS, A. J., WHITTINGHAM, M. J., BRADBURY, R. B., WILSON, J. D., KYRKOS, A., BUCKINGHAM, D. L. & EVANS, A. E. (2001). Foraging habitat selection by yellowhammers (Emberiza citrinella) nesting in agriculturally contrasting regions in lowland England. Biological Conservation 101, 197-210.

MORRISON, M. (2002). Searcher bias and scavenging rates in bird/wind energy studies. NREL/SR-500-30876.

MORRISON, M. L. (1998). Avian Risk and Fatality Protocol. NREL/SR-500-24997.

MORRISON, M. L. & POLLOCK, K. H. (1997). Development of a practical modelling framework for estimationg the impact of wind technology on bird populations. NREL/SR-440-23088.

MORRISON, M. L. & POLLOCK, K. H. (2000). Development of a Practical Modeling Framework for Estimating the Impact of Wind Technology on Bird Populations. In Proceedings of National Avian - Wind Power Planning Meeting III (ed. PNAWPPM-III), pp. 183-187. Prepared for the Avian Subcommittee of the National Wind Coordinating Committee by LGL Ltd., King City, Ont., San Diego, California.

MORRISON, M. L., POLLOCK, K. H., OBERG, A. L. & SINCLAIR, K. C. (1998). Predicting the response of bird populations to wind energy-related deaths. In Wind Energy Symposium. A collection of the 1998 ASME Wind Energy Symposium Technical Papers at the 36th AIAA Aerospace Sciences, pp. 157-164.

MÜLLER, A. & ILLNER, H. (2002). Beeinflussen Windenergieanalagen die Verteilung rufender Wachtelkönige und Wachteln? In Windenergie und Vögel - Ausmaß und Bewältigung eines Konfliktes, Technische Universität Berlin.

MUSTERS, C. J. M., NOORDERVLIET, M. A. W. & KEURS, W. J. T. (1996). Bird casualities caused by a wind energy project in an estuary. Bird Study 43, 124-126.

NABU (2004). Vögel der Agrarlandschaft - Bestand, Gefährdung, Schutz. NABU-Broschüre, 1-44.

NABU BAG WEISSSTORCHSCHUTZ (2004). Mitteilungsblatt 96/2004. Loburg: NABU BAG Weissstorchschutz, 20 S.

ORLOFF, S. & FLANNERY, A. (1992). Wind turbine effects on avian activity, habitat use and mortality in Altamont Pass and Solano County wind resources areas 1989-1991. California Energy Commission, Bio-Systems Analysis, Tiburon, Califonia.

ORLOFF, S. & FLANNERY, A. (1996). A continued examination of avian mortality in the Altamont Pass Wind Ressource Area. California Energy Commission, Sacramento; Bio-Systems Analysis, Inc., Santa Cruz, California.

OSBORN, R. G., HIGGINS, K. F., DIETER, C. D. & USGAARD, R. E. (1996). Bat collisions with wind turbines in Southwest Minnesota. Bat Research News 37, 105-108.

OWEN, M. & BLACK, J. M. (1989). Factors affecting the survival of Barnacle Geese on migration from the breeding grounds. Journal of Animal Ecology 34, 601-647.

PEACH, W. J., THOMPSON, P. S. & COULSON, J. C. (1994). Annual and long-term variation in the survival rates of British lapwings Vanellus vanellus. Journal of Animal Ecology 63, 60-70.

PEARCE-HIGGINS, J. W. & YALDEN, D. W. (2003). Golden Plover Pluvialis apricaria breeding success on a moor managed for shooting Red Grouse Lagopus lagopus. Bird Study 50, 170-177.

PEDERSEN, M. B. & POULSEN, E. (1991a). Impact of a 90m/2 MW wind turbine on birds. Avian responses to the implementation of the Tjaereborg Wind Turbine at the Danish Wadden Sea. Dansk Vildtundersogelser Kalø 47.

PEDERSEN, M.-B. & POULSEN, E. (1991b). En 90m/2 MW vindmoelles invirking pa fuglelivet. Fugles reaktioner pa opfoerelse og ideftsaettelsen af tjaereborgmoellen ved Det Danske Vadehav. Danske Vildundersoegelser 47, 44.

PERCIVAL, S. M. (2000). Birds and wind turbines in Britain. British Wildlife 12, 8-15.

PHILLIPS, J. F. (1994). The effects of a windfarm on the upland breeding bird communities of Bryn Titli, Mid Wales: 1993-1994. RSPB, The Welsh Office, Newtown.

POOT, H. (2004). Effects of artificial light of different colours on (nocturnally) migrating birds. Manuskript.

PRANGE, H. (1989). Der Graue Kranich. NBB 229, Ziemsen, Wittenberg Lutherstadt, 272 S.

PRÉVOT-JUILLARD, A.-C., LEBRETON, J.-D. & PRADEL, R. (1998). Re-evaluation of adult survival of Black-headed Gulls (Larus ridibundus) in presence of recapture heterogeneity. Auk 115, 85-95.

RAHMEL, U., BACH, L., BRINKMANN, R., LIMPENS, H. & ROSCHEN, A. (2004). Windenergieanlagen und Fledermäuse - Hinweise zur Erfassungsmethodik und zu planerischen Aspekten. Bremer Beiträge für Naturkunde und Naturschutz 1-12.

REICHENBACH, M. (2002). Windenergie und Wiesenvögel - wie empfindlich sind die Offenlandbrüter? In Windenergie und Vögel - Ausmaß und Bewältigung eines Konfliktes (ed. H. Ohlenburg), pp. 52-76. Technische Universität, Berlin.

REICHENBACH, M. (2003). Auswirkungen von Windenergieanlagen auf Vögel - Ausmaß und planerische Bewältigung, Technische Universität, Berlin.

REICHENBACH, M. & SCHADEK, U. (2003). Langzeituntersuchungen zum Konfliktthema

„Windkraft und Vögel". 2. Zwischenbericht. Unveröffentlichtes Gutachten im Auftrag des Bundesverbandes Windenergie.

REICHENBACH, M. & SINNING, F. (2003). Empfindlichkeiten ausgewählter Vogelarten gegenüber Windenergieanlagen - Ausmaß und planerische Bewältigung. In Kommen die Vögel und Fledermäuse unter die (Wind)räder?, Dresden, 17.-18.11.2003.

SACHS, L. (1978). Angewandte Statistik. 5. Aufl. Springer, Berlin, Heidelberg, New York.

SACHSLEHNER, L. & KOLLAR, H. P. (1997). Vogelschutz und Windkraftanlagen in Wien. Stadt Wien, Wien.

SAGE, R. B. & ROBINSON, P. A. (1996). Factors affecting songbird communities using new short rotation coppice habitats in spring. Bird Study 43, 201-213.

SALEK, M. (1993). Breeding of Lapwing (Vanellus vanellus) in Basins of South Bohemia: population density and habitat preference. Sylvia 30, 46-58.

SCHEKKERMAN, H. & MÜSKENS", G. (2000). Produceren Grutto's Limosa limosa in agrarisch grasland voldoende jongen voor en duurzame populatie? Limosa 73, 121-134.

SCHERNER, E. R. (1999). Windkraftanlagen und „wertgebende Vogelbestände" bei Bremerhaven: Realität oder Realsatire? Beiträge zur Naturkunde Niedersachsens 52, 121-156.

SCHMIDT, E., PIAGGIO, A. J., BOCK, C. E. & ARMSTRONG, D. M. (2003). National Wind Technology Center Site Environmental Assessment: Bird and Bat Use and Fatalities - Final Report; Period of Perfomance: April 23, 2001 - December 31, 2002. NREL/SR-500-32981.

SCHREIBER, M. (1992). Rastvögel und deren Habitatwahl im Bereich „Westermarsch" (Landkreis Aurich) im Jahr 1992. Unveröff. Gutachten im Auftrag der Ingenieursgemeinschaft agwa.

SCHREIBER, M. (1993a). Windkraftanlagen und Watvogel-Rastplätz. Naturschutz und Landschaftspflege 25, 133-139.

SCHREIBER, M. (1993b). Windkraftanlagen und Watvogel-Rastplätze - Störungen und Rastzplatzwahl von Brachvogel und Goldregenpfeifer. Naturschutz und Landschaftsplanung 25, 133-139.

SCHREIBER, M. (1993c). Zum Einfluß von Störungen auf die Rastplatzwahl von Watvögeln. Informationsd. Natursch. Nieders. 13, 161-169.

SCHREIBER, M. (1999). Windkraftanlagen als Störungsquelle für Gastvögel am Beispiel von Blessgans (Anser albifrons) und Lachmöwe (Larus ridibundus). Bremer Beiträge für Naturkunde und Naturschutz 4, 39-48.

SCHREIBER, M. (2000). Windkraftanlagen als Störquellen für Gastvögel. In Empfehlungen des Bundesamtes für Naturschutz zu naturschutzverträglichen Windkraftanlagen (ed. A. Winkelbrandt, R. Bless, M. Herbert, K. Kröger, T. Merck, B. Netz-Gerten, J. Schiller, S. Schubert and B. Schweppe-Kraft). Landwirtschaftsverlag, Münster.

SCHREIBER, M. (2002). Einfluss von Windenergieanlagen auf Rastvögel und Konsequenzen für EU-Vogelschutzgebiete. In Windenergie und Vögel - Ausmaß und Bewältigung eines Konfliktes (ed. H. Ohlenburg), pp. 134-156. Technische Universität Berlin, Berlin.

STIEFEL, A. & H. SCHEUFLER (1984). Der Rotschenkel. NBB 562, Ziemsen, Wittenberg Lutherstadt, 172 S.

STRUWE-JUHL, B. (1995). Auswirkungen der Renaturierungsmaßnahmen im Hohner See-Gebiet auf Bestand, Bruterfolg und Nahrungsökologie der Uferschnepfe (Limosa limosa). Corax 16, 153-172.

STRUWE-JUHL, B. (2002). Altersstruktur und Reproduktion des Seeadlerbrutbestands (Haliaeetus albicilla) in Schleswig-Holstein. Corax 19, Sonderheft 1, 51-61.

WILKENS, S. & EXO, K.-M. (1998). Brutbestand und Dichteabhängigkeit des Bruterfolges der Silbermöwe (Larus argentatus) auf Mellum. Journal für Ornitholgie 139, 21-37.

SELLIN, D. (1994). Notizen zum Vorkommen der Wachtel im Raum Wolfen-Zörbig. Apus 8, 265-270.

SEO, S. E. D. O. (1995). Effects of wind turbine power plants on the Avifauna in the Campo de Gibraltar region. Sociedad Espanola de Ornitologia SEO.

SGS ENVIRONMENT. (1994). Haverigg windfarm ornithological monitoring programme. Report to Windcluster LTD.

SINNING, F. (1999). Ergebnisse von Brut- und Rastvoguntersuchungen im Bereich des Jade-Windparks und DEWI-Testfelds in Wilhelmshaven. Bremer Beiträge für Naturkunde und Naturschutz 4, 61-70.

SINNING, F. & GERJETS, D. (1999). Untersuchungen zu Annäherung rastender Vögel in Windparks in Nordwestdeutschland. Bremer Beiträge für Naturkunde und Naturschutz 4, 53-59.

SMALLWOOD, K. S. & THELANDER, C. G. (2004). Developing methods to reduce bird mortality in the Altamont Pass Wind Ressource Area., pp. 1-363. Final report by BioResource Consultants to the California Energy Commission.

SOMMERHAGE, M. (1997). Verhaltensweisen ausgesuchter Vogelarten gegenüber Windkraftanlagen auf der Vaßbecker Hochfläche (Landkreis Waldeck-Frankenberg). Vogelkundliche Hefte Edertal 23, 104-109.

STEIOF, K., BECKER, J. & RATHGEBER, J. (2002). Ornithologische Stellungnahme zur Erweiterung der Windenergieanlage bei Mildenberg (Kreis Oberhavel, Land Brandenburg). Gutachten im Auftrag der Windenergie Wenger-Rosenau GmbH, Berlin.

STERNER, D. (2002). A roadmap for PIER research on avian collisions with wind turbines in California. California Energy Commission.

STIEBEL, H. (1997). Habitatwahl, Habitatnutzung und Bruterfolg der Schafstelze Motacilla flava in einer Agrarlandschaft. Vogelwelt 118, 257-268.

STILL, D., LITTLE, B. & LAWRENCE, E. S. (1996). The effect of wind turbines on the bird population at Blyth Harbour, Northumberland. ETSU W/13/00394/REP.

STRICKLAND, M. D., ERICKSON, W. P., JOHNSON, G., YOUNG, D. & GOOD, R. (2001a). Risk Reduction Avian Studies at the Foote Creek Rim Wind Plant in Wyoming. In Proceedings of National Avian - Wind Power Planning Meeting IV (ed. PNAWPPM-IV), pp. 107-114. Prepared for the Avian Subcommittee of the National Wind Coordinating Committee by RESOLVE, Inc., Washington, D.C., Susan Savitt Schwartz, Carmel, California.

STRICKLAND, M. D., JOHNSON, G., ERICKSON, W. P. & KRONNER, K. (2001b). Avian Studies at Wind Plants Located at Buffalo Ridge, Minnesota and Vansycle Ridge, Oregon. In Proceedings of National Avian - Wind Power Planning Meeting IV (ed. PNAWPPM-IV), pp. 38-52. Prepared for the Avian Subcommittee of the National Wind Coordinating Committee by RESOLVE, Inc., Washington, D.C., Susan Savitt Schwartz, Carmel, California.

STÜBING, S. & BOHLE, H. W. (2001). Untersuchungen zum Einfluss von Windenergieanlagen auf Brutvögel im Vogelsberg (Mittelhessen). Vogelkundliche Berichte aus Niedersachsen 33, 111-118.

SUTER, C., REHSTEINER, U. & ZBINDEN, N. (2002). Habitatwahl und Bruterfolg der Grauammer Miliaria calandra im Grossen Moos. Ornithologischer Beobachter 99, 105-115.

THELANDER, C. G. & RUGGE, L. (2000). Avian risk behavior and fatalities at the Altamont Wind Resource Area, March 1998 to February 1999. NREL/SR-500-27545.

THELANDER, C. G., SMALLWOOD, K. S. & RUGGE, L. (2003). Bird risk behaviors and fatalities at the Altamont Pass Wind Resource Area. Period of performance: March 1998 - December 2000. NREL/SR-500-33829.

TÖPFER, S. (1996). Beziehungen zwischen Landschaftsstruktur und Vogelbeständen einer Agrarlandschaft im nördlichen Harzvorland. Diplomarbeit, Universität Halle.

TRAPP, H., FABIAN, D., FÖRSTER, F. & ZINKE, O. (2002). Fledermausverluste in einem Windpark. Naturschutzarbeit in Sachsen 44, 53-56.

TRYJANOWSKI, P. & BAJCZYK, R. (1999). Population decline of the Yellow Wagtail Motacilla flava in an intensively used farmland of western Poland. Vogelwelt 120, Suppl., 205-207.

UGORETZ, S. (2001). Avian Mortalities at Tall Structures. In Proceedings of National Avian - Wind Power Planning Meeting IV (ed. PNAWPPM-IV), pp. 165-166. Prepared for the Avian Subcommittee of the National Wind Coordinating Committee by RESOLVE, Inc., Washington, D.C., Susan Savitt Schwartz, Carmel, California.

UMWELTBUNDESAMT. (2001). Wasserkraftanlagen als erneuerbare Energiequelle - rechtliche und ökologische Aspekte. UBA-Texte 01/01.

US FISH AND WILDLIFE SERVICE. (2003). Interim guidelines to avoid and minimize wildlife impacts from wind turbines. United States Department of the Interior, Fish and Wildlife Service, Washington, D.C.

VAN DER WINDEN, J., SPAANS, A. L. & DIRKSEN, S. (1999). Nocturnal collision risks of local wintering birds with wind turbines in wetlands. Bremer Beiträge für Naturkunde und Naturschutz 4, 33-38.

VERHOEF, J. P., WESTRA, C. A., KORTERINK, H. & CURVERS, A. WT-Bird. A novel bird impact detection system. ECN research Centre of the Netherlands.

VIERHAUS, H. (2000). Neues von unseren Fledermäusen. ABU Info 24, 58-60.

WAKEHAM-DAESON, A. & AEBISCHER, N. J. (1997). Arable reversion to permanent grassland: determining best management options to benefit declining grassland bird populations. Game Conservancy Trust, Fordingbridge.

WALTER, G. & BRUX, H. (1999). Erste Ergebnisse eines dreijährigen Brut- und Gastvogelmonitorings (1994-1997) im Einzugsbereich von zwei Windparks im Landkreis Cuxhaven. Bremer Beiträge für Naturkunde und Naturschutz 4, 81-106.

WEIBEL, U. (1999). Effects of wildflower strips in an intensively used arable area on skylarks (Alauda arvensis). Dissertation, Eidgenössische Technische Hochschule Zürich.

WEIBEL, U. M. (1995). Auswirkungen von Buntbrachen auf die Territorialität, Brutbiologie und Nahrungsökologie der Feldlerche Alauda arvensis. Diplomarbeit, Eidgenössische Technische Hochschule Zürich.

WILSON, J. D., EVANS, J., BROWN, S. J. & KING, J. R. (1997). Territory distribution and breeding success of skylarks Alauda arvensis on organic and intensive farmland in southern England. Journal of Applied Ecology 34, 1462-1478.

WINKELMAN, J. E. (1989). Vogels in het windpark nabij Urk (NOP): aanvaringsslachtoffers en verstoring van pleisternde eenden, ganzen en zwanen. RIN-rapport 89/15, Arnhem.

WINKELMAN, J. E. (1992a). De invloed van de Sepproef Windcentrale te Oosterbierum (Fr.) op vogels. 1: aanvaringsslachtoffers, pp. 71. DLO Instituut voor Bos- en Natuuronderzoek (Hrsg.), Arnhem.

WINKELMAN, J. E. (1992b). De invloed van de Sep-proefwindcentrale te Oosterbierum (Fr.) op vogels, 4: verstoring. RIN-rapport92/5, Arnhem.

YOUNG, D. P., ERICKSON, W. P., GOOD, R. E., STRICKLAND, M. D. & JOHNSON, G. D. (2003a). Avian and bat mortality associated with the initial phase of the Foote Creek Rim Windpower project, Carbon County, Wyoming. Final report. Western EcoSystems Technology, Inc., Wyoming.

YOUNG, D. P., ERICKSON, W. P., STRICKLAND, M. D., GOOD, R. E. & SERNKA, K. J. (2003b). Comparison of avian responses to UV-light-reflective paint on wind turbines. Western EcoSystems Technology, NREL/SR-500-32840, Cheyenne.

www.ingramcontent.com/pod-product-compliance
Lightning Source LLC
Chambersburg PA
CBHW080001230526
45470CB00008B/2816

The purpose of this report is to compile and to evaluate the available information on the impacts of exploitation of renewable energy sources on birds and bats. The focus is on wind energy as there is only little information on the impact on birds and bats of other sources of renewable energy.

The report aims at better understanding the size of the impact, the potential effects of re-powering (exchanging small old wind turbines by new big turbines), and possible measures to reduce the negative impact on birds by wind turbines. In addition the need for further research is highlighted.

ISBN 3-8334-5257-9